CHEMISTRY RESEARCH AND APPLICATIONS

SUPEROXIDE DISMUTASE

STRUCTURE, SYNTHESIS AND APPLICATIONS

Chemistry Research and Applications

Additional books in this series can be found on Nova's website under the Series tab.

Additional e-books in this series can be found on Nova's website under the eBooks tab.

CHEMISTRY RESEARCH AND APPLICATIONS

SUPEROXIDE DISMUTASE

STRUCTURE, SYNTHESIS AND APPLICATIONS

SERGEI MAGLIOZZI
EDITOR

nova
science publishers
New York

NOTICE TO THE READER

Library of Congress Cataloging-in-Publication Data

ISBN: 978-1-53613-080-5

Published by Nova Science Publishers, Inc. † New York

CONTENTS

PREFACE

In this compilation, the authors comprehensively investigate SOD effects in the homeostasis of mammal endometrium, using available information on several species and their team experience relating to the topic. In addition, they also address its role in endometrial integrity and some uterine clinical conditions and infertility. The current knowledge of plant SODs, their abiotic-stress modulated expression and activity, and analyses results on genetic engineering of plant SODs are summarized. Significance of superoxide dismutases in the crop improvement for stress tolerance is also discussed. This book reviews the oxidative stress and damage to plants, while also summarizing the characteristics of SOD enzymes and discussing their involvement in the tolerance of plants against abiotic stress. Additionally, many authors have studied the protective role of SOD in the mice cochlea, however more recently the role of SOD gene polymorphisms in the susceptibility of sudden sensorineural hearing loss has been investigated. Therefore, the authors also examine the role of SOD in the cochlea and its involvement in the pathogenesis of noise induced hearing loss, age related hearing loss and sudden sensorineural hearing loss.

Chapter 1 - Free radicals and other reactive oxygen species (ROS) are continuously generated as byproducts of normal cellular metabolism. Cells face up against ROS using a primary antioxidant system, characterized by a group of enzymes working together and sequentially, to reduce or eliminate ROS. The three major classes of antioxidant enzymes are the superoxide dismutase, catalase, and glutathione peroxidase. Superoxide dismutase (SOD) refers to a family of enzymes that catalyze the dismutation of superoxide to hydrogen peroxide and molecular oxygen. Superoxide and SOD are necessary to maintain biological homeostasis through various functions involving several cellular signal transduction pathways. Since the discovery of SOD more than four decades ago, there have been extensive studies on their biological activities in health and disease. The unbalanced activity of those enzymes (including SOD) plays a role in the pathogenesis of many diseases such as cancer, neurodegenerative diseases, allergy, ischemia/reperfusion injury or atherosclerosis. The endometrium is a very peculiar organ that in intact, mature females responds to the effects of cyclic alternance of sex steroids (estrogens and progesterone) to accomplish complex functions, including sperm tolerance, fertilization, embryo implantation and subsequent formation of the placenta. Cyclic changes in both ROS production and SOD activity have been reported to occur in the endometrium of different species. Moreover, it has been proposed that disturbances in ROS equilibrium might induce damages to tissues and predispose to infertility and many uterine diseases. Cyclical variations in the expression of superoxide dismutase (SOD) have been reported in the human and bitch endometria. Increased SOD activity has been related to proliferation in the follicular stage of the menstrual cycle and increased levels of estrogen. In contrast, SOD activity decreases in the late secretory phase. SOD activity has been related to angiogenesis during endometrial growth, regeneration, shedding or in implantation, due to its interplay with the local cytokine network. Also, SOD downregulation may trigger apoptosis, resulting from the accumulation of ROS in tissues. Furthermore, in the uterus, a decrease in SOD activity has been connected to prostaglandin F2-alpha synthesis and the endometrial secretion at progesterone withdrawal and luteolysis in species with short lifespan corpus luteum. Differences between species are expected in regards to SOD activity during the estrous or menstrual cycles and might be related with some physiological species-specificities concerning the endometrial cycle or the type and chronology of implantation, e.g., estrogen-mediated reduction in SOD activity, as described in the canine endometrium during the early luteal stage, may be associated with an increase in the membrane fluidity of endometrial cells and therefore be favorable to embryo invasion. In this chapter, the authors comprehensively investigate SOD effects in the homeostasis of mammal endometrium, using available information on several species and the authors' team experience in the topic. In addition, the authors will also address its role in endometrial integrity and some uterine clinical conditions and infertility.

Chapter 2 - Abiotic stresses affect plant growth and development, and significantly reduce crop productivity by causing the alteration in cell biochemistry. Common for all abiotic stresses is that they cause an imbalance between production and scavenging of reactive oxygen species (ROS) in cells, resulting in elevation of ROS concentrations to cytotoxic level. Superoxide dismutases (SODs) are enzymes that play an important role of primary ROS scavengers by catalyzing the conversion of superoxide anion radical to hydrogen peroxide and oxygen. Most plant species contain numerous SOD isoforms which are classified according to the active site metal into three major types: FeSOD, MnSOD and Cu/ZnSOD. All plant SODs are nuclear-encoded, synthesized on cytosolic ribosomes and transported into various cell compartments, such as cytosol, plastids, mitochondria, peroxisomes and cell wall. Expression of SODs genes might be modulated by developmental and environmental cues, including nutrition and abiotic stresses. So far, enhanced expression and/or activity of SODs have been reported in response to various abiotic stressors, such as water deficit, high salinity, heavy metals, hypoxia, high and low temperatures. This chapter summarizes the current knowledge of plant SODs, their abiotic-stress modulated expression and activity, and analyses results on genetic engineering of plant SODs. The significance of superoxide dismutases in the crop improvement for stress tolerance is also discussed.

Chapter 3 - Oxidative stress is induced by a wide range of abiotic stresses. Drought and salt stresses give usually birth to the generation of reacting oxygen species (ROS) such as hydrogen peroxide (H_2O_2) and superoxide ($O_2^{\bullet -}$). Those two species produced in a number of cellular reactions meaning various oxidoreductases such as lipoxygenases, peroxidases, NADPH oxidase and xanthine oxidase. To control the level of ROS under stress conditions, plants have involved a large scale of enzymatic and non-enzymatic antioxidants to scavenge ROS and to restore the redox homeostasis of the cells. Indeed, the superoxide dismutases (SODs) are very important attendants of the antioxidant biocatalysts due to the dismutation of O^{2-} to H_2O_2 and are located in the main cell compartments. In addition, Fe-SOD, Mn-SOD and Cu/Zn-SOD are the groups according to the heavy metal co-factor used by the enzyme and they often act in a contradictory manner after abiotic stress. In this chapter, the authors review the oxidative stress and damage to plants. Furthermore, the authors try to summarize the characteristics of SOD enzymes and discuss their involvement in the tolerance of plants against abiotic stress.

Chapter 4 - Superoxide dismutase (SOD) is an antioxidant agent that can be found in almost every cell exposed to oxygen. It is responsible for the dismutation of superoxide anions resulting from cellular metabolism into oxygen and hydrogen peroxide. In particular, superoxide dismutase (SOD), in its cytoplasmic isoform, contains copper (Cu) and zinc (Zn), and can be found in cochlear tissue (stria vascularis, spiral ligament, spiral ganglion neurons and organ of Corti). The antioxidant function of SOD is important for protection of the inner ear cells from toxic substances, such as reactive oxygen species (ROS). The role of SOD has been investigated in noise-induced hearing loss and in presbycusis, whereas elevated concentrations of ROS are responsible for inner ear damage. In the last decades, many authors have tested the protective role of SOD in the cochlea of animal models. Recently, the role of SOD gene polymorphisms in the susceptibility of sudden sensorineural hearing loss has been investigated as well. The aim of this chapter is to review the role of SOD in the cochlear metabolism and its involvement in the pathogenesis of noise induced hearing loss, age related hearing loss and sudden sensorineural hearing loss.

In: Superoxide Dismutase
Editor: Sergei Magliozzi
ISBN: 978-1-53613-080-5

Chapter 1

THE ROLE OF SUPEROXIDE DISMUTASE IN ENDOMETRIAL FUNCTIONAL INTEGRITY

Dario Santos*[1]*, PhD and Rita Payan-Carreira*[2]*, PhD
[1]CITAB & Department of Biology and Environment, Universidade de Trás-os-Montes e Alto Douro, Vila Real, Portugal
[2]CECAV & Department of Zootechnics Universidade de Trás-os-Montes e Alto Douro, Vila Real, Portugal

ABSTRACT

Free radicals and other reactive oxygen species (ROS) are continuously generated as byproducts of normal cellular metabolism. Cells face up against ROS using a primary antioxidant system, characterized by a group of enzymes working together and sequentially, to reduce or eliminate ROS. The three major classes of antioxidant enzymes are the superoxide dismutase, catalase, and glutathione peroxidase. Superoxide dismutase (SOD) refers to a family of enzymes that catalyze the dismutation of superoxide to hydrogen peroxide and molecular oxygen. Superoxide and SOD are necessary to maintain biological homeostasis through various functions involving several cellular signal transduction pathways. Since the discovery of SOD more than four decades ago, there have been extensive studies on their biological activities in health and disease. The unbalanced activity of those enzymes (including SOD) plays a role in the pathogenesis of many diseases such as cancer, neurodegenerative diseases, allergy, ischemia/reperfusion injury or atherosclerosis.

The endometrium is a very peculiar organ that in intact, mature females responds to the effects of cyclic alternance of sex steroids (estrogens and progesterone) to accomplish complex functions, including sperm tolerance, fertilization, embryo implantation and subsequent formation of the placenta. Cyclic changes in both ROS production and SOD activity have been reported to occur in the endometrium of different species. Moreover, it has been proposed that disturbances in ROS equilibrium might induce damages to tissues and predispose to infertility and many uterine diseases.

Cyclical variations in the expression of superoxide dismutase (SOD) have been reported in the human and bitch endometria. Increased SOD activity has been related to proliferation in the follicular stage of the menstrual cycle and increased levels of estrogen. In contrast, SOD activity decreases in the late secretory phase.

SOD activity has been related to angiogenesis during endometrial growth, regeneration, shedding or in implantation, due to its interplay with the local cytokine network. Also, SOD downregulation may trigger apoptosis, resulting from the accumulation of ROS in tissues. Furthermore, in the uterus, a decrease in SOD activity has been connected to prostaglandin F2-alpha synthesis and the endometrial secretion at progesterone withdrawal and luteolysis in species with short lifespan corpus luteum.

Differences between species are expected in regards to SOD activity during the estrous or menstrual cycles and might be related with some physiological species-specificities concerning the endometrial cycle or the type and chronology of implantation, e.g., estrogen-mediated reduction in SOD activity, as described in the canine endometrium during the early luteal stage, may be associated with an increase in the membrane fluidity of endometrial cells and therefore be favorable to embryo invasion.

In this chapter, we comprehensively investigate SOD effects in the homeostasis of mammal endometrium, using available information on several species and our team experience in the topic. In addition, we will also address its role in endometrial integrity and some uterine clinical conditions and infertility.

Keywords: superoxide dismutase, reactive oxygen species, oxidative stress, fertility, diseases, mammal endometrium

INTRODUCTION

Free radicals and other reactive oxygen species (ROS) are continuously generated as byproducts of normal cellular metabolism (Gate et al. 1999). These ROS are necessary to maintain biological homeostasis through various functions involving several cellular signal transduction pathways (Thannickal & Fanburg 2000; Finkel, 2011; Zhang et al., 2016). However, an excessive production of ROS, an impaired antioxidant system or a combination of these factors, results in oxidative stress (OS) that cause damage to the macromolecules necessary for cell structure and function (Dröge, 2002). Cells face up against ROS using antioxidants, defined as any substance that delays, prevents, or removes oxidative damage to a target molecule (Halliwell & Gutterdige 2015). The primary antioxidants systems are characterized by a group of enzymes working together and sequentially to reduce or eliminate ROS (Halliwell, 2007). The three major classes of antioxidant enzymes are the superoxide dismutase (SOD), catalase and glutathione peroxidase (GPx) (Valko et al. 2007). Superoxide dismutase refers to a family of enzymes that catalyze superoxide elimination in the first reaction of a free radical removal pathway (Fukai & Fukai, 2011)

The free radical superoxide was first described by L. Pauling in the 1930's (Pauling, 1979). Superoxide synthesis by the enzyme xanthine oxidase was firstly studied by Knowles and coworkers in 1969 (Knowles et al., 1969), and during the following decades several biologically occurring reactions that produce substantial amounts of O_2- were found (Fridovitch, 1978a; Fridovitch, 1978b; Halliwell & Gutteridge 1984). Enzymatic superoxide dismutation by *Cu,ZnSOD (SOD1* in mammals) was identified by McCord and Fridovich in 1969 (McCord & Fridovich, 1969), despite that the purification of the enzyme has been achieved 30 years earlier, from bovine blood and liver, as a copper binding protein with an unknown function (Mann & Keil, 1938). In the early 1970s, Fridovich and coworkers further discovered *MnSOD* (*SOD2* in mammals) and *FeSOD* (not present in mammals) (Keele et al., 1970; Yost & Fridovich, 1973). Finally, the third mammal SOD isoenzyme (*ECDOD, SOD3* in mammals) was discovered by Marklund and coworkers in 1982 (Marklund et al., 1982). Recently, a novel prokaryotic SOD isoenzyme has been discovered that relies on nickel (*NiSOD*) (Barondeau et al., 2004; Wuerges et al., 2004).

Table 1. Characteristics of mammalian superoxide dismutase isoenzymes

Isoenzymes	SOD1	SOD2	SOD3
	Cu,ZnSOD - Copper, zinc superoxide dismutase	MnSOD – Manganese superoxide dismutase	ECSOD – Extracellular superoxide dismutase
Cellular location	Cytoplasm, lysosomes, peroxisomes, mitochondrial intermembrane space and nuclear compartments	Mitochondrial matrix	Extracellular space
Molecular mass	32 kDa	86-88 kDa	135 kDa
Assembly and molecular mass of subunits	Homodimer (16 kDa monomers with 153 aa)	Homotetramer (22 - kDa subunit)	Homotetramer (30 - kDa subunits)
Metal ions	1 Cu + 1 Zn[(1)]	Mn[(2)]	1 Cu + 1 Zn[(3)]
Chromosomal localization	21q22	6q25	4q21
References	Keller et al., 1991; Crapo et al., 1992; Liou, 1993; Halliwell & Gutteridge, 2015	Halliwell & Gutteridge, 2015	Fattman et al., 2003

(1) Cu interacts with histidine residues at the SOD active site, and Zn stabilizes the enzyme.

(2) Mn at the enzyme active center.

(3) SOD3 contains a high affinity domain for heparin, which contributes to its localization in the extracellular matrix.

The SOD1 isoform is present in the cytosol, liposomes, nucleus and intermembrane mitochondrial space and is a homodimer with a molecular mass of 32 kDa. Both SOD2 and SOD3 are homotetramers with a molecular mass of 86-88 and 135 kDa and are localized in the mitochondrial matrix and extracellular space, respectively. Enzymatic activity of SOD1 and SOD3 depends on the presence of copper (Cu) and zinc (Zn) (Fridovich, 1975). Zn participates in the proper protein folding and stability while Cu directly participates in SOD catalytic activity (Fukai & Fukai, 2011; McCord & Fridovich, 1975). Manganese (Mn) at the active site of SOD2 serves to catalyze the dismutation of superoxide similarly to Cu in SOD1 and SOD3 (Hsu et al., 1996). Table 1 summarizes the general characteristics of the three mammalian SOD isoforms.

Superoxide and superoxide oxidase studies remain the focus of several studies in biology and health sciences so that 'the discovery and naming of SOD are, in the opinion of many researchers, the most important discovery of modern biology never to win the Nobel Prize' (Lane, 2002).

General Characteristics

Even though the mammal isoenzymes of SOD – SOD1, SOD2, and SOD3 – differ in amino acid sequence, they catalyze essentially the same biochemical reaction (Zelko, 2002). Superoxide dismutase (EC 1.15.1.1) catalyzes the dismutation of superoxide radical according to the reaction:

$$2O_2^- + 2H^+ \rightarrow H_2O_2 + O_2$$

Dismutase refers to a kind of chemical reaction in which the same reactant is both oxidized and reduced (Cammack, 2006). SOD-catalyzed dismutation is characterized by the oxidation of one superoxide anion to molecular oxygen and the reduction of another to hydrogen peroxide (Halliwell & Gutteridge, 2015). Superoxide dismutation process is nearly independent of pH, occurring within the pH range of 5.0 to 9.5. It occurs by a successive oxidation and reduction of a transition metal ion, such as copper (Cu) and manganese (Mn), at the active site of the enzyme (Abreu & Cabelli, 2010; Fridovich, 1975), in a "ping-pong" mechanism (Figure 1), with a high reaction rate of $1.5 \times 10^9\ M^{-1}s^{-1}$ (Fridovich, 1975).

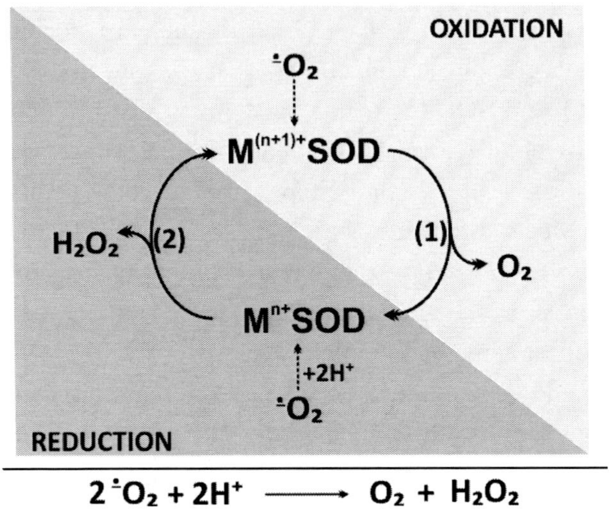

Figure 1. SOD-catalyzed superoxide dismutation. **M** (**M**etal) = copper (**Cu**; n = 1); manganese (**Mn**; n = 2); iron (**Fe**; n = 2); and nickel (**Ni**; n = 2). Oxidation (1) and reduction (2) dismutation half-reactions are presented in two distinct grey triangles. The overall reaction is written below.

All the SOD isoenzymes bind differentially monovalent anions (e.g., azide, fluoride), with SOD 1 being competitively inhibited in the presence of Fl^-, N_3^-, and CN^- (Leone et al., 1998; Meier et al., 1998). The three mammalian SOD isoenzymes have been identified with a distinct pattern of tissue distribution (Zelko, 2002). In humans, the liver has a relatively high amount and activity of SOD1, while the activity of SOD2 is higher in the renal cortex than in other tissues, and SOD3 has been identified in plasma and lymph (Halliwell & Gutteridge, 2015).

Several transcription factors are involved in SOD cellular control (Miao and St. Clair, 2009). Superoxide can induce the expression of several genes constitutively regulating an inducible expression of SOD genes (Birben et al., 2012). Within these redox-sensitive transcription factors are included Nrf2/Keap1, NF-κB, activator proteins (AP-1, AP-2), specificity protein (SP1) and CCAA/enhancer binding protein (C/EBP) (Miao & St. Clair, 2009; Sies et al. 2017). A post-transcriptional regulation of SOD expression includes changes in mRNA stability, mRNA translation and protein post-translational modification (Yamakura & Kawasaki, 2010).

SOD Role in Health and Disease

Several animal studies evidenced a crucial role for SOD isoenzymes in protecting against a wide variety of diseases. The substantial number of evidence supporting a causal involvement of superoxide in multiple disease processes came from studies in animal models (Noor et al., 2002). These studies showed that transgenic overexpression of SOD inhibits disease processes, while an opposite effect may be obtained with a genetic deletion of SOD (Li et al. 1995; Melov et al., 1998; Shen et al., 2006; Valko et al., 2007). Researchers still look for the role of superoxide and SOD in human disease pathogenesis (Halliwell & Gutteridge, 2015). Changes in superoxide dismutase expression or content in several organs or systems are related to the pathogenesis and pathophysiology of diverse conditions (Hopkins, 2016).

Aging

Changes in SOD activities correlated with the age of animals (McCord, 2002 Melov et al., 2000; Warner, 1994). Lifespan extension has been achieved in *Caenorhabditis elegans* and *Drosophila melanogaster* by an induced over-expression of SOD (Orr & Sohal, 1994; Sohal et al., 1995) or the use of enzymatic synthetic drugs mimicking the enzyme activity (Kregel & Zhang, 2007). However, several studies failed to detect significant lifespan increase with improved antioxidant defenses (including overexpression of SOD1 and SOD2) in vertebrate species (Huang et al., 2000; Van Remmen et al., 2003; Page et al., 2010; Salway et al., 2010).

Cancer

Many studies with gene knockout animals showed that reduction of SOD1 or SOD2 results in increased DNA damage, mutations and a higher incidence of cancer development (Van Remmen et al., 2003; Elchuri et al.,

2005). Conversely, overexpression of SOD3 inhibits breast carcinoma cell growth and invasion (Kensler et al., 1983; Teoh et al., 2009). SOD levels are lower in cancer cells or higher in tumor cell lines that are resistant to drug treatments (Kim et al., 2005; Sandoval et al., 2006; Hurt et al., 2007; Park et al., 2009).

Immunological Disorders

Oxidative stress due to SOD1 deficiency accelerates the destruction of red blood cells, leading to anemia. Antibodies produced against the oxidation products may trigger the autoimmune reaction (Iuchi et al. 2009). The expression of SOD2 is decreased in HIV-infected patients, followed by increased oxidative stress with enhanced protein carbonylation and lipid peroxidation (Flores et al., 1993). In contrast, the expression of cytosolic SOD1 in macrophages is increased during HIV infection *in vitro*, most likely to counteract elevated superoxide anion concentrations (Delmas-Beauvieux et al., 1996). SOD-1 plays a significant role in immune processes, because modulation of its activity may differentially affect the NO-dependent microbicidal activity and release of cytokines by activated macrophages (Marikovsky et al., 2003).

Neurodegeneration

Compelling evidence associates the oxidative stress and SOD activity with neurodegenerative diseases, such as amyotrophic lateral sclerosis (Deng et al. 1993; Valentine et al. 2005; Kabuta et al., 2009; Bergemalm et al., 2010), Parkinson´s diseases, Huntington´s disease, Alzheimer´s disease (Massaad et al., 2009; Murakami et al., 2011) and other degenerative disorders. Mutations in the SOD1 gene are known to play a role in at least 20% of familial amyotrophic lateral sclerosis cases (Rosen et al., 1993; Perry et al., 2010).

Pulmonary Disorders

A beneficial role for all three isozymes of SOD in pulmonary hyperoxic injury has been demonstrated in experimental animals via both gene knockout and transgenic overexpression (Gao et al., 2008; Kinula & Crapo, 2009). SOD attenuate pulmonary inflammatory responses, oxidative stress, and tissue remodeling following hyperoxia (Folz et al., 1999; Yao et al., 2010). Overexpression of SODs also results in amelioration of pulmonary injury induced by drugs, radiation, and environmental chemicals or dust (Kinula & Crapo, 2009). Targeted disruption of SODs aggravates the pathophysiological processes of the pulmonary disorders (St. Clair et al., 1991). Diminished lung SOD activity in patients with pulmonary arterial hypertension (PAH) indicates an impaired capacity to detoxify superoxide radical (Bowers et al., 2004), confirmed by the observation that extracellular SOD overexpression in lungs was able to ameliorate PAH induced by monocrotaline (Kamezaki et al., 2008).

Cardiovascular Diseases

Gene knockout and transgenic overexpression animal studies demonstrate a crucial role of SOD isoenzymes in cardiovascular disorders, including myocardial ischemia-reperfusion injury, atherosclerosis, hypertension, heart failure, cardiac hypertrophy, and drug/xenobiotic-induced cardiovascular complications (Yen et al., 1996; Levonen et al., 2001). Deletion of SOD2 in mice is embryonically lethal or causes early neonatal death due primarily to dilated cardiomyopathy (Li et al., 1995; Lebovitz et al., 1996). SOD3 activity is increased in macrophage-rich atherosclerotic lesions of apolipoprotein E-deficient mice (Fukai et al., 1998). In contrast, the enzymatic activity of SOD3 was decreased in atherosclerotic lesions of the human aorta as compared with normal aorta segments of the same individual (Luoma et al., 1998). Vascular SOD3 activity was reduced while activities of SOD1 and SOD2 were similar in the coronary arteries of patients with coronary artery disease and control subjects (Landmesser et al., 2000). After short periods of ischemia, reoxygenation of isolated perfused animal hearts has been shown to produce additional tissue damage that is diminished by adding SOD1 or SOD2 to the reperfusion fluids. In addition, hearts from transgenic animals overexpressing SOD2 or SOD3 show less reperfusion injury, whereas animals lacking such antioxidant enzymes show more severe heart injuries (Levonen et al., 2008; Richters et al., 2011; Obal et al., 2012).

Hepatic and Gastrointestinal Diseases

Several animal studies demonstrate a protective role for all three SOD isoenzymes in hepatic disorders (Laukkanen et al., 2001; Wheeler et al., 2001), including graft dysfunction after liver transplantation, nonalcoholic fatty liver disease, liver ischemia-reperfusion injury, and liver injury induced by alcohol and other drugs/xenobiotics. Homozygous SOD2-deficient mice died within a month of life with a cardiac disorder and increased concentration of lipids in the liver (Li et al., 1995). Viral vector-mediated delivery of either SOD1 or SOD2, but not SOD3 increases animal survival and attenuates graft dysfunction after transplantation of fatty livers in rats. It has also been reported that SOD1 deficiency increases hepatic lipid accumulation in mice (Uchiyama et al., 2006). Overexpression of either SOD1 or SOD2 in mice inhibits alcohol-induced liver injury, whereas knockout of SOD1 aggravates the liver damage (Wheeler et al., 2001; Kessova et al., 2003). Overexpression of SOD1 also protects against carbon tetrachloride-elicited hepatotoxicity in mice. However, mice deficient in SOD1 are resistance to acetaminophen-induced hepatotoxicity, suggesting a possible pro-oxidant role for the physiological level of SOD1 activity in acetaminophen-mediated liver injury (Lei et al., 2006). A transgenic overexpression of SOD1 in mice protects tissue from neutrophil infiltration and lipid peroxidation during intestinal ischemia-reperfusion. Oral co-administration of SOD1 and catalase is reported to attenuate stress-induced gastric mucosal lesions in rats. In an animal model of chemically-induced acute colitis, transgenic overexpression of SOD1 results in amelioration of colonic inflammation and improves animal survival (Kruidenier et al., 2003).

Diabetes

Several experimental findings suggest a critical role for SOD1 and SOD2 in protecting against diabetes development and various diabetes complications (Kubisch et al., 1994; Halliwell & Gutteridge, 2015). Overexpression of SOD1 or SOD2 attenuates diabetic retinopathy, neuropathy, renal injury and cardiomyopathy (Kowluru et al., 2006; Shen et al., 2006; Vincent et al., 2007). The overexpression of SOD1 protects against end-organ damage in models of type II diabetic nephropathy (DeRubertis et al., 2007). Studies in mice with genetic deletions of various antioxidant enzymes have also provided insight into the specific relative contributions of SOD2 to the development of diabetes complications (Hinerfeld et al., 2004). Further strengthening a potential role for the antioxidant SOD2, specific polymorphisms in the SOD2 gene were associated with the development of diabetic complications (Mollsten et al., 2004).

Renal Diseases

In transplanted human kidneys that had been subjected to cold storage, SOD infusion into the renal artery at implantation produced significant improvement in renal function (Pollak et al., 1993). Either viral vector-mediated or transgenic overexpression of SODs is reported to attenuate renal ischemia-reperfusion injury in animal models. Targeting a synthetic cationic SOD to renal proximal tubular cells inhibits oxidative stress and nephrotoxicity of cisplatin (an anticancer drug that may cause severe renal injury) and increases the survival of cancer-bearing mice (Nishikawa et al., 2001). SOD1 deficiency causes salt sensitivity and aggravates hypertension in hydronephrosis in mice (Carlstrom et al., 2009). Similarly, mice with heterozygous knockout of SOD2 (SOD2+/-) developed salt-sensitive hypertension and accelerated renal senescence (Rodriguez-Iturbe et al., 2007). Viral delivery of SOD1 gene reduces cyclosporine A (an immunosuppressive agent used in organ transplantation but is nephrotoxic) induced free radical formation and nephrotoxicity.

Reproductive Disorders

In the testes, endogenous SOD protects spermatozoa from excessive ROS (Tsunoda et al., 2012). SOD was shown to preserve sperm motility and reduce lipid peroxidation. Sperm incubation with high concentrations of SOD preserved motility in mouse spermatozoa and enhanced mice embryo development (Nonogaki et al., 1992). Another study, evaluating the effects of SOD on sperm viability, motility, and morphology, found that SOD was associated with improved semen quality (Perumal, 2014). A higher proportion of sperm exposed to superoxide anion exhibited capacitation and hyperactivation, as opposed to those not exposed to SOD (de Lamirande et al., 1997). Moreover, the gene transfer of SOD3 to the penis reduces superoxide levels and improves erectile function in aged or diabetic rats (Bivalacqua et al., 2003; Deng et al. 2010). An increased level of oxidative stress associated with a decreased SOD activity was associated with the polycystic ovarian syndrome. It may play a pathogenetic role in the development of insulin resistance (Suresh et al., 2015; Kucukaydinet et al., 2016; Victor et al., 2016), hyperandrogenism (Gonzalez et al., 2012; Suresh et al. 2015), and chronic inflammation observed in this disease (Papalou et al., 2012; Seleem et al., 2014; Pertynska-Marczewska et al., 2015; Zuo et al., 2016).

Other Diseases and Conditions

In septic patients, production of superoxide anion is increased, and its levels in plasma are elevated compared to healthy normal (Warner et al.,

1995). Increased levels of SOD were also reported in a pediatric septic population, although no prognostic correlation with outcome was observed (Batra et al., 2000). Oxidative stress is involved in the pathogenesis of several skin disorders, including ultraviolet light-induced injury, skin carcinogenesis, and skin aging (Murakami et al. 2009). Treatment with SOD1 ameliorates ischemic skin injury in experimental animals (Galenko-Yaroshevskii et al., 2006). Levels of SOD are increased in proliferating keratinocytes of injured mouse skin, and mice lacking SOD1 show impaired wound healing (Taylor et al., 1993; Bennaars-Eiden et al., 2002). Activated leukocytes release Inflammation mediators including cytokines, eicosanoids, and nitric oxide (NO). Excess NO can react with superoxide to form peroxynitrite anion, a potent ROS. It has been reported that overproduction of tumor necrosis factor (TNF)-α, which is mainly released by macrophages, is associated with a decrease in SOD1 expression, whereas TNF-α, IL-1, IL-4, and IL-6 activate SOD2 (Kiningham et al., 2001; Afonso et al., 2006). Interferon-γ plus TNF-α exposure has been shown to increase SOD3 secretion in dermal fibroblasts and SOD3 expression along with inducible NO synthase in rat type II pneumocytes (Marklund, 1992; Brady et al., 1997). SODs may be essential protectors in anemia. Deficiency of either SOD1 or SOD2 in mice causes oxidative stress in red blood cells and hemolytic anemia (Friedman et al., 2004; Iuchi et al., 2009). Studies also demonstrate a protective role for SODs in aging-related cognitive decline and muscle atrophy (Muller et al., 2006).

SOD Role in the Endometrial Function

The mammalian endometrium is an utterly complex structure that undergoes finely tuned morphological changes in response to ovarian sex steroids, with the ultimate goal of guaranteeing embryo survival, implantation, and the success of pregnancy (Ponnampalam et al., 2004; Sherwin et al., 2006). These cyclic changes encompass a species-specific pattern of proliferation, differentiation, and remodeling, which is supposedly related to the species physiology, namely the characteristics of ovarian cycle, litter size and the placental type (Payan-Carreira et al., 2016).

Like any other highly dynamic tissue, the metabolism of endometrial cells continuously generates reactive oxygen species. Various studies identified ROS and several antioxidant enzymes in the mammal endometrium. It was also demonstrated that OS modulates the cyclic changes in the endometrium (Gupta et al., 2008). ROS are now acknowledged as crucial for female fertility (Rizk et al., 2013). ROS homeostatic concentrations are kept in equilibrium by a pro-oxidant/antioxidant balance, whose disruption would predispose to disease (Burton & Jauniaux, 2011).

SOD has been identified in the endometrium in human (Narimoto et al., 1990; Sugino et al., 2004; Sugino, 2007), sheep (Al-Gubory et al., 2008, 2017), cows (Giergiel & Kankofer, 2014), guinea pigs (Makker et al., 2006) and dogs (Payan-Carreira et al., 2016), as well as in uterine cell preparations in mice (Jain et al., 1999). Cyclic hormonal changes across the endometrial cycle modulate the local antioxidant activity, since ROS and antioxidant enzymes respond to sex steroid stimulation, as it was demonstrated by Serviddio et al. (2002).

Sugino′ studies showed that SOD is highly expressed in the glandular epithelia and stroma of the human endometrium (Sugino et al., 1996; Sugino et al., 2000; Sugino, 2007). Both SOD and ROS were evidenced in the endometrium. The changes in SOD and ROS described throughout the endometrial cycle were proposed to support a critical role in the regulation of endometrial function.

SOD Role in the Cyclic Endometrium

The immunolocalization of SOD in the human endometrium showed that SOD content increase from the proliferative to the secretory phase, peaking in mid- to late-secretory phases (Narimoto et al., 1990; Sugiro et al., 1996). A similar pattern of expression was found by Payan-Carreira et al. (2016) in the canine endometrium. There, SOD immunolabeling was higher in all the epithelia compared to the stroma. Also, the intensity of the reaction decreased from anestrus (low estrogen and progesterone levels) to proestrus (rise in estrogens), and then increased from estrus (transition from estrogen to progesterone dominance) to early and mid diestrus (luteal phase) (Payan-Carreira et al., 2016). Furthermore, differences in the pattern of SOD immunoexpression were described in the different epithelial elements of the canine endometrium (surface epithelium and superficial and deep glandular epithelia), which could be due to the distinct involvement and function of these structures in the canine implantation and placentation.

SOD was also detected in the endometrial fluid of humans (Rahiminejad et al., 2016), cows (Ramos et al., 2015) and dogs (Kobayashi et al., 2014). Rahiminejad et al. (2016) found higher SOD levels in endometrial secretions in women with ongoing pregnancies than in those that failed IVF cycles, but no differences were detected among women with successful pregnancies and miscarriage. Ramos et al. (2015) reported that, in cows, SOD activities in uterine washings were not affected by the periovulatory sex steroid milieu and suggested that this could be related to a dilution effect in the secretions that minimized any putative changes. In contrast, these authors reported an increased expression of SOD before ovulation, which might contribute to cell survival. In dog uterine fluid, SOD activity was higher in diestrus compared with anestrus or estrus (Kobayashi et al., 2014), which according to the authors would improve sperm quality and fertility at breeding. It is important to notice that in dogs, ovulation occurs under progesterone levels around 5ng/mL, due to pre-ovulatory luteinization of granulosa cells (Concannon, 2011), while in the cow it occurs in an estrogenic environment (Ramos et al., 2015).

Enzymatic assays testing SOD activity in explants or extracts of endometrial tissue in humans showed that SOD activity rises from early to the mid-late proliferative stage, transiently decreasing during the early secretory and resuming high activity levels in the middle secretory stage and early pregnancy (Sugino et al., 1996). In dogs, total SOD activity (Figure 2) was found higher in anestrus (basal estrogen and progesterone levels) and proestrus (rising estrogen levels) but it was inhibited by increasing levels of progesterone (progesterone levels start rising in dogs in estrus –when estrogen levels fall - and remain above basal levels through the early and mid diestrus) (R. Payan-Carreira & D. Santos, unpublished data). Conversely, this pattern in SOD activity seems to contradict the previous Immunohistochemistry (IHC) study (Payan-Carreira et al., 2016). Still, the differences could be explained by the fact that in the IHC study, only the protein expression of SOD 1 was identified in the canine uterus, and it was not limited to the activated enzyme. In mice, estrogen reduces the total SOD activity in the uterus, which could favor the endometrial plasticity of endometrial cells and favor embryo interaction (Jain et al., 1999).

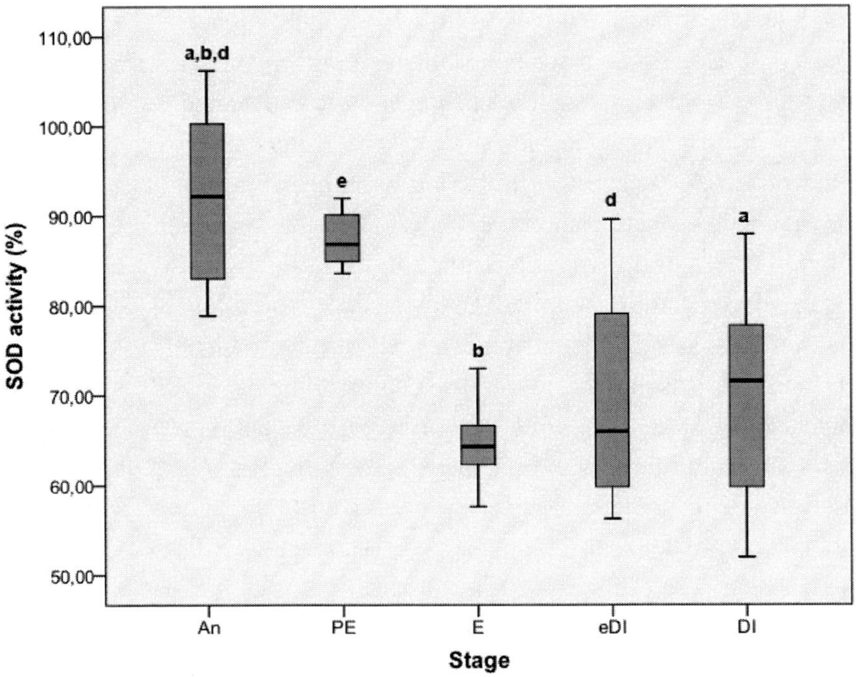

Figure 2. Box-plot graph for superoxide dismutase (SOD) activity in the different stages of the canine estrous cycle: anestrus (A), proestrus (PE), estrus (E), early diestrus (eDI) and diestrus (DI). Five animals were used per group. Values sharing the same letter are significantly different: $P<0.05$ (a/d/e); $P<0.01$ (b).

It has been shown that SOD, and thereby ROS in physiological concentrations, mediates hormone signaling, apoptosis, cell migration, angiogenesis, and prostaglandin secretion (Sugino et al., 2001; Agarwal et al., 2005; Sugiro, 2007), and is of primordial importance in embryo implantation and decidualization (Sugino et al., 2001; Rizk et al., 2013).

In humans, SOD content decreases in the late secretory phase preceding menstruation (Sugiro et al., 1996), which has been associated with a decrease in both estrogen and progesterone stimulation. The decrease in SOD activity is followed by an increase in local ROS. These, in turn, would stimulate prostaglandin F2alpha secretion via activation of nuclear factor kappa B and consequently command endometrial shedding (Sugino et al., 2001; Wu et al., 2014), as well as luteolysis (Du Plessis et al., 2017). A decrease in SOD activity was not detected in the canine endometrium (Figure 2), but this may be related to the species particularities, such as the low levels of PGF2alpha recorded at the end of the secretory phase (Luz et al., 2006), and the existence of a prolonged lifespan corpora lutea (in dogs, cyclic and gestational corpora lutea are of similar lifespan – Concannon, 2011). Besides, the activation of the COX2 enzyme will also stimulate pro-inflammatory cytokines like TNF (Sugino, 2007; Agarwal et al., 2016).

SOD1 interacts with several cytokine pathways mediated by TNF-α (Afonso et al., 2006) or the transforming growth factor (TGF)-beta (Valko et al., 2007), through which it may interfere with both cell apoptosis and proliferation pathways, respectively. Under an increased production of ROS, TNF activates Caspase 3, thereby stimulating apoptosis via the extrinsic pathway (Figure 3) (Sugino, 2007; Agarwal et al., 2016). It is accepted that an increase in SOD activity will protect cells from TNF pro-apoptotic effects (Sugiro, 2007; Burton & Jauniaux, 2011). However, it is important to remember that SOD activity - and thereby, ROS - may also mediate cell apoptosis in the uterus through different mechanisms (Figure 3) (Leitão et al., 2010; Burton & Jauniaux, 2011), using both the intrinsic and the extrinsic pathways. The activation of one or the other pathway is context-dependent and therefore may vary with the stage of the cycle or the reproductive status of the female, possibly depending on the activation of different concurrent regulatory pathways.

In addition, SOD has been associated with cell proliferation and tissue remodeling, by controlling the activity of metalloproteinases 2 or 9 through the modulation of the local concentration of ROS (Fukai et al., 2011).

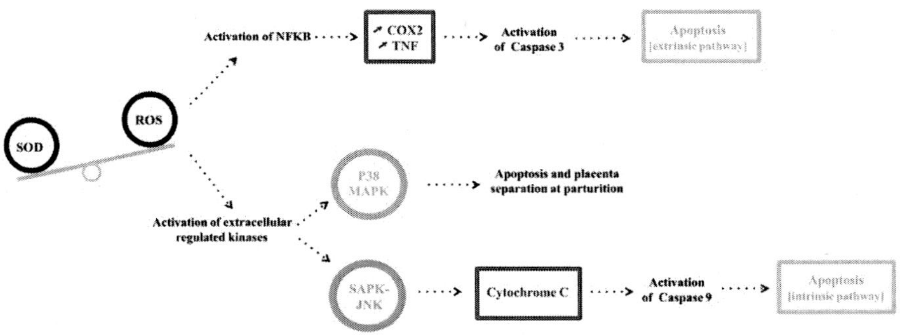

Figure 3. Superoxide dismutase (SOD) involvement in the pathways regulating cell apoptosis.

Remodeling of the vascular network occurs in all stages of the endometrial cycle, under a tightly regulated process; these changes provide the necessary support to cell growth and differentiation in the endometrium. Local temporary hypoxia, associated with decreased SOD activity, is a major regulator of the endometrial angiogenesis in the endometrium. It also plays a crucial role in embryo implantation and formation of the placenta in mammals (Okada et al., 2016). In general, during the proliferative phase, the growth of new vessels should accompany endometrial proliferation and promote the regulation of vascular permeability, while in the secretory phase the endometrium possesses a well-developed vascular network (Smith, 2001) that supports the glandular secretory activity and foster receptivity towards any embryo. In this aspect, induction of angiogenesis occurs similarly to the observed in wound healing and neoplasia, and in close relation with a disruption in circulation induced by hypoxia, altogether with a rapid proliferation and growth characteristic of these states (Marikovsky et al., 2002). Available information raises the hypothesis that different pathways are involved in the regulation of endometrial angiogenesis besides the activation of SOD enzyme. Some studies showed that hypoxia and ROS are involved in the regulation of endometrial microvasculature. It can be achieved by the vascular endothelial growth factor (VEGF, proangiogenic factor) and Angiopoietin-2 (Ang-2, angiogenic antagonist) (Agarwal et al., 2016); VEGF activity is modulated via TNF and TGFbeta1 pathways (Okada et al., 2016). An increase in SOD activities, along with the derived hypoxia, around the time of implantation serves to trigger endometrial angiogenesis, via VEGF activation (Du Plessis et al., 2017), under a very precise mechanism.

SOD Role in Implantation and Decidualization

Placenta formation follows the embryo implantation. If the embryo implantation follows a more or less similar mechanism across species, despite its different chronologies, the placentation varies greatly between species, particularly respecting the invasion of the maternal endometrium and the gross placenta morphology, which in turn originates differences in the degree of modification of the endometrial layers and of the vascular network supporting the placenta (Carter & Enders, 2016).

In animals developing a decidua, after the apposition phase at implantation, the embryonic trophoblast invades the endometrium, triggering a more or less important, species-specific, tissue remodeling that is accompanied by the proliferation of stromal and glandular epithelial cells, as well as by changes in the local vascular network. The occurrence of oxidative stress at the embryo-maternal interface starts on early pregnancy and, particularly in this stage, ROS play a dual role in early pregnancy events. Its homeostatic regulation is decisive to the outcome of pregnancy (Burton & Jauniaux, 2011).

Early embryonic development and placentation occur in a state of low oxygen. Modest hypoxia favors trophoblast differentiation and its invasive ability toward the maternal endometrium (Sugino, et al., 1996; Sugino, 2007; Rizk et al., 2013). In mice, a decrease in SOD activity, and a subsequent sharp increase in superoxide anion radical was observed at the time of implantation (Nivsarkar et al., 2005).

The initial differentiation of the decidual cells, located below the surface epithelia in humans (Kajihara et al., 2016) and dogs (Payan-Carreira et al., 2014) occurs even in the absence of an embryo, at the mid-late luteal phase of the cycle (Kajihara et al., 2016). In the pregnant endometrium, this preliminary decidualization is sustained by the progesterone secretion. The withdraw of progesterone receptors observed in non-pregnant cycles at the end of the luteal stage is reversed by the decrease in SOD activity that is recorded at mid to late secretory phases in pregnant cycles (Sugino, 2007; Kajihara et al., 2016; Du Plessis et al., 2017), impairing the release of prostaglandin F2alpha that would trigger luteolysis.

It has also been demonstrated that decidual cells are more resistant to oxidative cell death than non-decidualized stromal cells (Kajihara et al., 2016). This resistance is associated with the inhibition of the c-Jun NH-terminal kinase (JNK) stress signaling (intrinsic) pathway by SOD2 (Kajihara et al., 2016). Inhibition of SOD2 led to increased ROS production, triggering apoptosis in rabbit endothelial cells (Liu et al. 2011). The inactivation of this pathway will also contribute to the maintenance of the progesterone receptor activity in decidual cells (Leitão et al., 2010).

In sheep, around the moment of embryo attachment, an increase in the total SOD activity has been reported, particularly in the caruncular areas of the endometrium, which was mainly due to an increased in SOD 2 activity (Al-Gubory et al., 2016; Al-Gubory et al., 2017). The authors hypothesized that a switch in the activities of SOD 1 and 2 compared with those reported at the end of the luteal phase occurs in response to conceptus derived factors (Al-Gubory et al., 2016).

It has been shown that, in humans, the formation of placental villi and the subsequent formation of the placental disc is mediated by the local oxidative stress, which also regulates the remodeling of the endometrial spiral arteries (Burton & Jauniaux, 2011). Simultaneously with villi formation, the endometrium goes through an intensive remodeling of its vascular network to support embryo development and growth (Burton &

Jauniaux, 2011). This remodeling occurs on both the maternal and fetal sides of the placenta. It results from either the elongation of pre-existing endothelial tubes (non-branching angiogenesis) or the extension of those tubes (spouting angiogenesis) (Shaman et al., 2013). It may also be originated by vasculogenesis (development of new vessels from mesenchymal stem cells) (Pereira et al., 2015), which starts on the trophoblast side (Torry et al., 2007). These changes in the endometrial vascular network are mediated by the trophoblast, which usually reverses the physiological pathways associated with luteolysis and progesterone withdraw, namely by inhibiting the NF-kB pathway, which up-regulates VEGF and modulates TNF and other cytokines (e.g., Interleukin 6 and 8 and TGF-beta 1) (Pereira et al., 2015). The homeostasis of the system is finely tuned by diverse molecules, including the Hypoxia Inducible Factor (HIF).

SOD Involvement in the Endometrial Pathology

Across species, disruption of the oxidative stress balance has been associated with endometrial diseases that can originate in cyclic or pregnant endometria (Rizk et al., 2013). However, little is known about the involvement of ROS or SOD in mammals' uterine diseases other than human or rodents.

Endometriosis is an important disease in human medicine, often leading to chronic pain and abdominal surgery, as well as infertility. It derives from the implantation of endometrial tissue within the pelvis, possibly favored by retrograde menstruation (Gupta & Agarwal, 2014). Although the results from different studies give contradictory information, it is now generally agreed that patients with endometriosis show an increase in the generation of ROS by activated peritoneal macrophages (Agarwal et al., 2006), derived from the decreased activity of enzymatic antioxidants in the peritoneal fluid, that together with increased lipid peroxidation, promote the growth and adhesion of endometrial cells in the peritoneal cavity (Sharma et al., 2017). This generates an unfavorable balance between antioxidants, particularly SOD, and ROS, originating cell damage, disturbed angiogenesis and bleeding, and persistence of inflammation.

A specific mutation in SOD2 enzyme has been associated with unexplained infertility in women (Gupta et al., 2014; Agarwal et al., 2016), impairing the role of some protein pathways that could lead to the pathophysiology of female infertility. Interference with the matrix metalloproteinases or with SOD2 regulated angiogenesis at placentation has been mentioned (Gupta et al., 2014), among other putative adverse effects at the level of ovaries and tuba.

Increased SOD activity has been reported in endometrial curettage samples of patients diagnosed with endometrial adenomyosis and with endometrial polyps when compared with endometrial hyperplasia and endometrial adenocarcinoma, the latter showing the lowest activities; however, in this study, the authors did not compare the enzyme activity with normal controls. The authors suggest that women with common benign gynecological diseases are more susceptible to develop precancerous lesions and cancer since the increased oxidative stress would stimulate cell cycle progression and promote cell proliferation.

Long-term progestin contraceptive treatment may induce disturbed endometrial angiogenesis with concurrent abnormal endometrial bleeding, which has been associated with ROS imbalance (Hickey et al., 2006).

Disturbed angiogenesis at embryo attachment has also been pointed out as one reason for idiopathic recurrent pregnancy loss (Aziz, 2013), which would originate endothelial damage, impaired placental vascularization and immune malfunction (Gupta et al., 2007). Aziz (2013) refers to a significant reduction in SOD activities when pregnancy loss occurs during the first trimester, while the staining intensity in trophoblast samples was higher in women that miscarriage before the day 77 compared with controls (Gupta et al., 2007).

Pre-eclampsia, a human pregnancy-specific condition, is a complex heterogeneous syndrome characterized by hypertension and proteinuria. Even though its exact etiology remains obscure, it is thought that may be originated on disrupted maternal endothelial function, which is probably related to a progressive ROS and SOD disequilibrium that will interfere with the normal vascular support to the placenta (Jauniaux & Burton, 2016). Sharma et al. (2006) showed that SOD circulating levels are elevated in women with pre-eclampsia, compared to matched control pregnancies. However, other syndromes associated with placental insufficiency and embryo growth retardation may develop in women without leading to pre-eclampsia. In the case of placental insufficiency, the endothelial dysfunction may be established during early pregnancy, when the development of placental vessels might be compromised by multiple causes (e.g., chromosomic abnormalities, chronic infections, immune-mediated pathologies, or trophoblast abnormalities) (Jauniaux & Burton, 2016).

CONCLUSION

SOD, as one of the primary antioxidant enzymes acting in oxidative stress, is involved in multiple and complex of regulation of physiologic processes, some of which occurring in the endometrium. ROS are kept in tightly controlled levels in tissues that guarantee the homeostasis of the organ. Excessive or deficient production of reactive oxygen species and/or a deficiency or excessive activities in the antioxidative defenses in the body triggers the disease.

Modest fluctuations in oxidative stress are allowed as part of the homeostatic process that is itself tightly regulated through the interaction with other regulatory pathways. This may contribute to the still limited amount of knowledge that is available on SOD and ROS participation in health and disease, despite the numerous studies that already contributed to explore the involvement of oxidative stress in female uterine diseases and infertility and putatively relevant treatment options.

We hope that this chapter provides the researchers with a concise reference on SOD and ROS involvement in endometrial physiology and medicine.

ACKNOWLEDGMENTS

We are grateful to Dr. Gustavo Paixão for the English language review.

The authors would like to acknowledge funding from the Portuguese Science and Technology Foundation (FCT) under the Projects PEst-OE/AGR/UI0772/2011 and PEst-OE/AGR/UI0772/2014.

REFERENCES

Abreu, I.A. & Cabelli, D, E. (2010). Superoxide dismutases-a review of the metal-associated mechanistic variations. *Biochim Biophys Ac*ta, 1804, 263-274.

Afonso, V., Santos, G., Collin, P., Khatib, A.M., Mitrovic, D.R., Lomri, N., Leitman, D.C. & Lomri, A. (2006). Tumor necrosis factor-[alpha] down - regulates human Cu/Zn superoxide dismutase 1 promoter via JNK/AP - 1 signaling pathway. *Free Radic Biol Med,* 41(5), 709-721.

Agarwal, A., Gupta, S. & Sharma, R.K. (2005). Role of oxidative stress in female reproduction. *Reprod Biol Endocrinol,* 3, 28.

Agarwal, A., Gupta, S., & Sikka, S. (2006). The role of free radicals and antioxidants in reproduction. *Curr Opin Obstet Gynecol*, 18(3), 325-332.

Agarwal, A., Tadros, H., Panicker, A., & Tvrdá, E. (2016). Role of oxidants and antioxidants in female reproduction. In *Oxidative Stress and Antioxidant Protection: The Science of Free Radical Biology and Disease*. Armstrong, D. and Stratton, R.D. (eds.). John Wiley and Sons. New York, NY. 253-280.

Al-Gubory, K.H., Bolifraud, P. & Garrel, C. (2008). Regulation of key antioxidant enzymatic systems in the sheep endometrium by ovarian steroids. *Endocrinology,* 149, 4428-4434.

Al-Gubory, K.H., Garrel, C., Sugino, N., & Fowler, P.A. (2016). The conceptus induces a switch in protein expression and activities of superoxide dismutase 1 and 2 in the sheep endometrium during early pregnancy. *Small Rumin Res*, 141, 77-83.

Al-Gubory, K.H., Faure, P., & Garrel, C. (2017). Different enzymatic antioxidative pathways operate within the sheep caruncular and intercaruncular endometrium throughout the estrous cycle and early pregnancy. *Theriogenology*, 99, 111-118.

Aziz, N. (2013). Recurrent Pregnancy Loss and Oxidative Stress. In *Studies on Women's Health*. Agarwal, A., Aziz, N. and Rizk, B. (eds.). Humana Press. Totowa, NJ, USA. 131-141.

Barondeau, D.P., Kassmann, C.J. & Bruns C.K. (2004). Nickel superoxide dismutase structure and mechanism. *Biochemistry,* 43, 8038-8047.

Batra, S., Kumar, R., Kapoor, A.K. and Ray, G. (2000) Alterations in antioxidant status during neonatal sepsis*, Ann Trop Paediatr,* 20(1), 27-33.

Bennaars-Eiden, A., Higgins, L., Hertzel, A.V., Kapphahn, R.J., Ferrington, D.A. & Bernlohr, D.A. (2002). Covalent modification of epithelial fatty acid-binding protein by 4-HNE *in vitro* and *in vivo*. *J Biol Chem*, 277, 50693-50702.

Bergemalm, D., Forsberg, K., Srivastava, V., Graffmo, K.S., Andersen, P.M., Brannstrom, T., Wingsle, G. & Marklund, S.L. (2010). Superoxide dismutase-1 and other proteins in inclusions from transgenic amyotrophic lateral sclerosis model mice. *J Neurochem,* 114(2), 408-418.

Bivalacqua, T.J., Armstrong, J.S., Biggerstaff, J., Abdel-Mageed, A.B., Kadowitz, P.J., Hellstrom, W.J. & Champion, H.C. (2003). Gene transfer of extracellular SOD to the penis reduces O2-. And improves erectile function in aged rats. *Am J Physiol Heart Circ Physiol,* 284, H1408-H1421.

Birben, E., Sahiner, U.M., Sackesen, C., Erzurum, S. & Kalayci, O. (2012). Oxidative stress and antioxidant defense. *World Allergy Organ J,* 5(1), 9-19.

Bowers, R., Cool, C., Murphy, R.C. Tuder, R.M., Hopken, M.W., Flores, S.C. & Voelkel, N.F. (2004). Oxidative stress in severe pulmonary hypertension. *Am J Respir Crit Care Med,* 169,764-769.

Brady, T.C., Chang, L.Y., Day, B.J. & Crapo, J.D. (1997). Extracellular superoxide dismutase is upregulated with inducible nitric oxide synthase after NF-kappa B activation. *Am J Physiol,* 273(5), L1002-L1006.

Burton, G.J. & Jauniaux, E. (2011). Oxidative stress. *Best Pract Res Clin Obstet Gynaecol*, 25(3), 287-299.

Cammack, R. (2006). *Oxford Dictionary of Biochemistry and Molecular Biology* (2006), 2nd edition, Oxford University Press, Oxford.

Carlstrom, M., Brown, R.D., Sallstrom, J., Larsson, E., Zilmer, M., Zabihi, S. Eriksson, U.J. & Persson, A.E. (2009). SOD1 deficiency causes salt sensitivity and aggravates hypertension in hydronephrosis. *Am J Physiol Regul Integr Comp Physiol*, 297, R82-92.

Carter, A.M., & Enders, A.C. (2016). Placentation in mammals: Definitive placenta, yolk sac, and paraplacenta. *Theriogenology*, 86(1), 278-287.

Concannon, P.W. (2011). Reproductive cycles of the domestic bitch. *Anim Reprod Sci*, 124, 200-210.

Crapo, J.D., Oury, T., Rabouille, C., Slot, J.W. & Chang, L.Y. (1992). Copper, zinc superoxide dismutase is primarily a cytosolic protein in human cells. *Proc Natl Acad Sci USA,* 89, 10405-10409.

Da Broi, M.G., de Albuquerque, F.O., de Andrade, A.Z., Cardoso, R.L., Jordao Junior, A.A. & Navarro, P.A. (2016). Increased concentration of 8-hydroxy-20-deoxyguanosine in follicular fluid of infertile women with endometriosis. *Cell Tissue Res,* 366(1), 231-242.

De Lamirande, E., Jiang, H., Zini, A., Kodama, H. & Gagnon, C. (1997). Reactive oxygen species and sperm physiology. *Rev Reprod,* 2(1), 48-54.

Delmas-Beauvieux, M.C., Peuchant, E., Couchouron, A., Constans, J., Sergeant, C., Simonoff, M, Pellegrin, J.L., Leng, B., Conri, C. & Clerc, M. (1996). The enzymatic antioxidant system in blood and glutathione status in human immunodeficiency virus (HIV)-infected patients: effects of supplementation with selenium or beta-carotene. *Am J Clin Nutr,* 64, 101-107.

Deng, H-X., Hentati, A., Tainer, J.A., Iqbal, Z., Cayabyab, A., Hung, W-Y., Getzoff, E.D., Hu, P., Herzfeldt, B., Roos, R.P., Warner, C., Deng,

G., Smyth, C., Parge, H.E., Ahmed, A., Roses, A.D., Hallewell, R.A., Pericak-Vance, M.A. & Siddique, T. (1993). Amyotrophic lateral sclerosis and structural defects in Cu, Zn superoxide dismutase. *Science,* 261(5124), 1047-1051.

Deng, W, Bivalacqua, T.J., Champion, H.C., Hellstrom, W.J., Murthy, S.N. & Kadowitz, P.J. (2010). Superoxide dismutase: a target for gene therapeutic approach to reduce oxidative stress in erectile dysfunction. *Methods Mol Biol*, 610, 213-227.

DeRubertis, F.R., Craven, P.A., Melhem, M.F. & Salah, E.M. (2004). Attenuation of renal injury in db/db mice overexpressing superoxide dismutase: evidence for reduced superoxide– nitric oxide interaction. *Diabetes,* 53(3), 762-768.

DeRubertis, F.R., Craven, P.A. & Melhem, M.F. (2007). Acceleration of diabetic renal injury in the superoxide dismutase knockout mouse: effects of tempol. *Metab Clin Exp,* 56, 1256-1264.

Dröge, W. (2002). Free radicals in the physiological control of cell function. *Physiol Review*, 82, 47-95.

Donnez, J., Binda, M.M., Donnez, O. & Dolmans, M.M. (2016). Oxidative stress in the pelvic cavity and its role in the pathogenesis of endometriosis. *Fertil Steril,* 106(5), 1011-1017.

Du Plessis, S.S., Harlev, A., Mohamed, M.I., Habib, E., Kothandaraman, N. & Cakar, Z. (2017). Physiological Roles of Reactive Oxygen Species (ROS) in the Reproductive System. In *Oxidative Stress in Human Reproduction*. Agarwal, A., Sharma, R., Gupta, S., Harlev, A., Ahmad, G., du Plessis, S.S., Esteves, S.C., Wang, S.W. and Durairajanayagam, D. (eds.) Springer International Publishing. 47-64.

Elchuri, S., Oberley, T.D., Qi, W., Eisenstein, R.S., Jackson Roberts, L., Van Remmen, H., Epstein, C.J. & Huang, T.T. (2005). CuZnSOD deficiency leads to persistent and widespread oxidative damage and hepatocarcinogenesis later in life. *Oncogene,* 24, 367-380.

Fattman, C.L., Schaefer, L.M. & Oury, T.D. (2003). Extracellular superoxide dismutase in biology and medicine. *Free Radic Biol Med,* 35, 236-256.

Finkel, T. (2011). Signal transduction by reactive oxygen species. *J Cell Biol*, *194*: 7-15.

Fukai, T., Galis, Z.S., Meng, X.P., Parthasarathy, S. & Harrison, D.G. (1998). Vascular expression of extracellular superoxide dismutase in atherosclerosis. *J Clin Invest* 101(10), 2101-2111.

Flores, S.C., Marecki, J.C., Harper, K.P., Bose, S.K., Nelson, S.K. & McCord, J.M. (1993). Tat protein of human immunodeficiency virus type 1 represses expression of manganese superoxide dismutase in HeLa cells. *Proc Natl Acad Sci USA,* 90, 7632-7636.

Friedman, J.S., Lopez, M.F., Fleming, M.D., Rivera, A., Martin, F.M., Welsh, M.L., Boyd, A., Doctrow, S.R. & Burakoff, S.J. (2004). SOD2-deficiency anemia: protein oxidation and altered protein expression reveal targets of damage, stress response, and antioxidant responsiveness. *Blood,* 104, 2565-2573.

Fridovich, I. (1975). Superoxide dismutases. *Ann Rev Biochem,* 44, 147-159.

Fridovich, I. (1978a). The biology of oxygen radicals. *Science*, 201, 875-880.

Fridovich, I. (1978b). Superoxide radicals, superoxide dismutase and the aerobic lifestyle. *Photochem Photobiol, 28*, 733-741.

Folz, R.J., Abushamaa, A.M. & Suliman, H.B. (1999). Extracellular superoxide dismutase in the airways of transgenic mice reduces inflammation and attenuates lung toxicity following hyperoxia. *J Clin Invest,* 103(7), 1055-1066.

Fukai T. & Ushio-Fukai, M. (2011). Superoxide dismutases: role in redox signaling, vascular function, and diseases. *Antioxid Redox Signal,* 15, 1583-1606.

Galenko-Yaroshevskii, V.P., Galka, V.V., Stoyalova, O.N., Agadzhanova, A.V., Lapina, I.V., Gaivoronskaya, T.V., Popkova, L.V., Churilova, I.V., Gorelashvili, A.S., Antelava, N.A., Berberashvili, T.M. & Sukoyan, G.V. (2006). Antinecrotic and antioxidant effects of superoxide dismutase during skin ischemia. *Bull Exp Biol Med,* 142: 447-449.

Gao, F., Kinnula, V.L., Myllarniemi, M. & Oury, T.D. (2008). Extracellular superoxide dismutase in pulmonary fibrosis. *Antioxid Redox Signal,* 10, 343-354.

Gate, L., Paul, J., Ba, G.N., Tew, K.D. & Tapiero, H. (1999). Oxidative stress induced in pathologies: the role of antioxidants. *Biomed Pharmacother,* 53, 169-180.

Giergiel, M., & Kankofer, M. (2014). The age-related changes in SOD activity in bovine (*Bos taurus*) reproductive organs. *Acta Sci Vet,* 42(1), 1241.

Gonzalez, F., Nair, K.S., Daniels, J.K., Basal, E., Schimke, J.M. & Blair, H.E. (2012). Hyperandrogenism sensitizes leukocytes to hyperglycemia to promote oxidative stress in lean reproductive-age women. *J Clin Endocrinol Metab,* 97(8), 2836-2843.

Gupta, S., Agarwal, A., Banerjee, J., & Alvarez, J.G. (2007). The role of oxidative stress in spontaneous abortion and recurrent pregnancy loss: a systematic review. *Obstet Gynecol Surv*, 62(5), 335-347.

Gupta, S. Sekhon, L., Aziz, N., & Agarwal, A. (2008). The Impact of Oxidative Stress on Female Reproduction and ART: Evidence-Based Review. In: *Infertility and Assisted Reproduction*. Makrigiannakis, A., Rizk, B., Velasco, R.G., and Sallam, H. (eds.). New York, NY: Cambridge University Press. 178-186.

Gupta, S., Ghulmiyyah, J., Sharma, R., Halabi, J. & Agarwal, A. (2014). Power of proteomics in linking oxidative stress and female infertility. *Biomed Res Int*, 2014.

Halliwell, B. (2007). Biochemistry of oxidative stress. *Biochem Soc Trans*. 35, 1147-1150.

Halliwell, B. & Gutteridge, J.M. (1984). Oxygen toxicity, oxygen radicals, transition metals and disease. *Biochem J, 219*, 1-14.

Halliwell, B. & Gutteridge, J.M.C. (2015). *Free radicals in biology and medicine.* 5th edition, Oxford University Press.

Hensley, K. & Floyd, R.A. (2002). Reactive oxygen species and protein oxidation in aging: a look back, a look ahead. *Arch Biochem Biophys,* 397, 377-383.

Hevir, N., Ribic-Pucelj, M. & Lanisnik, R.T. (2013). Disturbed balance between phase I and II metabolizing enzymes in ovarian endometriosis: a source of excessive hydroxy-estrogens and ROS? *Mol Cell Endocrinol,* 367(1-2), 74-84.

Hickey, M., Krikun, G., Kodaman, P., Schatz, F., Carati, C. & Lockwood, C.J. (2006). Long-term progestin-only contraceptives result in reduced endometrial blood flow and oxidative stress. *J Clin Endocrinol Metab*, 91(9), 3633-3638.

Hinerfeld, D., Traini, M.D., Weinberger, R.P., Bruce Cochran, B., Doctrow, S.R., Harry, J. & Melov, S (2004). Endogenous mitochondrial oxidative stress: neurodegeneration, proteomic analysis, specific respiratory chain defects, and efficacious antioxidant therapy in superoxide dismutase 2 null mice. *J Neurochem,* 88, 657-667.

Hopkins, R.Z. (2016). Superoxide in Biology and Medicine: An Overview. *Reactive. Oxygen Species* 1(2), 99-109.

Hsu, J.L., Hsieh, Y., Tu, C., O'Connor, D., Nick, H.S. & Silverman, D.N. (1996). Catalytic properties of human manganese superoxide dismutase. *J Biol Chem,* 271, 17687-17691.

Huang, T.T., Carlson, E.J., Gillespie, A.M., Shi, Y. & Epstein, C.J. (2000). Ubiquitous overexpression of CuZn superoxide dismutase does not extend life span in mice. *J. Gerontol A Biol Sci Med Sci,* 55(1), B5-B9.

Hurt, E.M., Thomas, S.B., Peng, B. & Farrar, W.L. (2007). Integrated molecular profiling of SOD2 expression in multiple myeloma. *Blood,* 109(9), 3953-3962.

Iuchi, Y., Okada, F., Takamiya, R., Kibe, N., Tsunoda, S., Nakajima, O., Toyoda, K., Nagae, R., Suematsu, M., Soga, T., Uchida, K. & Fujii, J. (2009). Rescue of anaemia and autoimmune responses in SOD1-deficient mice by transgenic expression of human SOD1 in erythrocytes. *Biochem J,* 422(2), 313-320.

Jain, S., Saxena, D., Kumar, P.G., Koide, S.S. & Laloraya, M. (1999). Effect of estradiol and selected antiestrogens on pro- and antioxidant pathways in mammalian uterus. *Contraception*, 60, 111-118.

Jauniaux, E. & Burton, G.J. (2016). Le rôle du stress oxydant dans les pathologies placentaires de la grossesse. *J Gynecol Obstet Biol Reprod (Paris)*, 45(8), 775-785.

Kabuta, T., Kinugawa, A., Tsuchiya, Y., Kabuta, C., Setsuie, R., Tateno, M., Araki, T. & Wada, K. (2009). Familial amyotrophic lateral sclerosis-linked mutant SOD1 aberrantly interacts with tubulin. *Biochem Biophys Res Commun,* 387, 121-126.

Kajihara, T., Ishihara, O. & Brosens, J.J. (2016). Oxidative Stress and Its Implications in Endometrial Function. In *Uterine Endometrial Function*. Kanzaki, H. (ed.) Springer, Japan.105-123.

Kamezaki, F., Tasaki, H., Yamashita, K., Tsutsui, M., Koide, S., Nakata, S., Tanimoto, A., Okazaki, M., Sasaguri, Y., Adachi, T. & Otsuji, Y. (2008). Gene transfer of extracellular superoxide dismutase ameliorates pulmonary hypertension in rats. *Am J Respir Crit Care Med,* 177, 219-226.

Keele, B.B. Jr, McCord, J.M. & Fridovich, I. (1970). Superoxide dismutase from *Escherichia coli* B. A new manganese-containing enzyme. *J Biol Chem,* 245, 6176-6181.

Keller, G.A., Warner, T.G., Steimer, K.S. & Hallewell, R.A. (1991). Cu, Zn superoxide dismutase is a peroxisomal enzyme in human fibroblasts and hepatoma cells. *Proc Natl Acad Sci USA,* 88, 7381-7385.

Kensler, T.W., Bush, D.M. & Kozumbo, W.J. (1983). Inhibition of tumor promotion by a biomimetic superoxide dismutase. *Science,* 221(4605), 75-77.

Kessova, I.G., Ho, Y.S., Thung, S. & Cederbaum, A.I. (2003). Alcohol-induced liver injury in mice lacking Cu, Zn-superoxide dismutase. *Hepatology,* 38, 1136-1145.

Kim, S.H., Kim, M.O., Gao, P., Youm, C.A., Park, H.R., Lee, T.S., Kim, K.S., Suh, J.G., Lee, H.T., Park, B.J., Ryoo, Z.Y. & Lee, T.H. (2005). Overexpression of extracellular superoxide dismutase (EC-SOD) in mouse skin plays a protective role in DMBA/TPA-induced tumor formation. *Oncol Res,* 15(7–8), 333-341.

Kiningham, K.K., Xu, Y., Daosukho, C., Popova, B. & St. Clair, D.K. (2001). Nuclear factor kappaB -dependent mechanisms coordinate the synergistic effect of PMA and cytokines on the induction of superoxide dismutase 2. *Biochem J,* 353 (1), 147-156.

Kinnula, V.L. & Crapo, J.D. (2003). Superoxide dismutases in the lung and human lung diseases. *Am J Respir Crit Care Med,* 167, 1600-1619.

Knowles, P.F., Gibson, J.F., Pick, F.M. & Bray, R.C. (1969). Electron-spin-resonance evidence for enzymic reduction of oxygen to a free radical, the superoxide ion. *Biochem J,* 111, 53-58.

Kobayashi, M., Wada, M., Hori, T. & Kawakami, E. (2014). Superoxide dismutase activity in the oviductal and uterine fluid of the bitch and the effects of the enzyme on viability, motility and hyperactivation of canine sperm in vitro. *J Vet Med Sci,* 76, 741-743.

Kowluru, R.A., Kowluru, V., Xion, Y. & Ho, Y.S. (2006). Overexpression of mitochondrial superoxide dismutase in mice protects the retina from diabetes-induced oxidative stress. *Free Radic Biol Med,* 41, 1191-1196.

Kregel, K.C. & Zhang, H.J. (2007). An integrated view of oxidative stress in aging: basic mechanisms, functional effects, and pathological considerations. *Am J Physiol Regul Integr Comp Physiol,* 292(1), 18-36.

Kruidenier, L., van Meeteren, M.E., Kuiper, I., Jaarsma, D., Lamers, C.B., Zijlstra, F.J. & Verspaget, H.W. (2003). Attenuated mild colonic inflammation and improved survival from severe DSS-colitis of transgenic Cu/Zn-SOD mice. *Free Radic Biol Med*, 34, 753-765.

Kubisch, H.M., Wang, J., Luche, R., Carlson, E., Bray, T.M., Epstein, C.J., & Phillips, J.P. (1994). Transgenic copper/zinc superoxide dismutase modulates susceptibility to type I diabetes. *Proc Natl Acad Sci USA,* 91(21), 9956-9959.

Kucukaydin, Z., Duran, C., Basaran, M., Camlica, F., Erdem, S.S., Basaran, A., Kutlu, O., Burnik, F.S., Elmas, H. & Gonen, M.S. (2016). Plasma total oxidant and antioxidant status after oral glucose tolerance and mixed meal tests in patients with polycystic ovary syndrome. *J Endocrinol Invest,* 39(10), 1139-1148.

Kukai, T. & Kukai, M.U. (2011). Superoxide Dismutases: Role in Redox Signaling, Vascular Function, and Diseases. *Antioxid Redox Signal,* 15(6), 1583–1606.

Landmesser, U., Merten, R., Spiekermann, S., Buttner, K., Drexler, H. & Hornig, B. (2000). Vascular Extracellular Superoxide Dismutase Activity in Patients with Coronary Artery Disease: Relation to Endothelium - Dependent Vasodilation. *Circulation,* 101(19), 2264-2270.

Lane, N. (2002). *Oxygen: the molecule that made the world.* Oxford University Press, New York.

Lei, X.G., Zhu, J.H., McClung, J.P., Aregullin, M. & Roneker, C.A. (2006). Mice deficient in Cu, Zn superoxide dismutase are resistant to acetaminophen toxicity. *Biochem J,* 399, 455-461.

Li, Y., Huang, T.T., Carlson, E.J., Melov, S., Ursell, P.C., Olson, J.L., Noble, L.J., Yoshimura, M.P., Berger, C., Chan, P.H., Wallace, D.C. & Epstein, C.J. (1995). Dilated cardiomyopathy and neonatal lethality in mutant mice lacking manganese superoxide dismutase. *Nat Genet,* 11, 376-381.

Laukkanen, M.O., Leppanen, P., Turunen, P., Tuomisto, T., Naarala, J. & Yla-Herttuala, S. (2001). EC-SOD gene therapy reduces paracetamol-induced liver damage in mice. *J Gene Med,* 3(4), 321-325.

Lebovitz, R.M., Zhang, H., Vogel, H., Cartwright, J., Jr., Dionne, L., Lu, N., Huang, S. & Matzuk, M.M. (1996). Neurodegeneration, myocardial injury, and perinatal death in mitochondrial superoxide dismutase-deficient mice. *Proc Natl Acad Sci USA,* 93(18), 9782-9787.

Leitao, B., Jones, M.C., Fusi, L., Higham, J., Lee, Y., Takano, M., Goto, T., Christian, M., Lam, E.W. & Brosens, J.J. (2010). Silencing of the JNK pathway maintains progesterone receptor activity in decidualizing human endometrial stromal cells exposed to oxidative stress signals. *FASEB J,* 24(5), 1541-1551.

Leone, M., Cupane, A., Militello, V., Stroppolo, M.E. & Desideri, A. (1998). Fourier Transform infrared analysis of the interaction of azide with the active site of oxidized and reduced bovine Cu, Zn superoxide dismutase. *Biochemistry,* 37, 4459-4464.

Levonen, A.L., Vahakangas, E., Koponen, J.K. & Yla-Herttuala, S. (2008). Antioxidant gene therapy for cardiovascular disease: current status and future perspectives. *Circulation,* 117, 2142-2150.

Li, Y., Huang TT, Carlson, E.J., Melov, S., Ursell, P.C., Olson, J.L., Noble, L.J., Yoshimura, M.P., Berger, C., Chan, P.H., Wallace, D.C. & Epstein, C.J. (1995). Dilated cardiomyopathy and neonatal lethality in mutant mice lacking manganese superoxide dismutase. *Nat Genet,* 11(4), 376-381.

Liou, W., Chang, L.Y., Geuze, H. J, S Crapo, J.D. & Slot, J.W. (1993). Distribution of CuZn superoxide dismutase in rat liver. *Free Radic Biol Med,* 14, 201-207.

Liu, C., Ogando, D. & Bonanno, J.A. (2011). SOD2 contributes to anti-oxidative capacity in rabbit corneal endothelial cells. *Mol Vis*, 17, 2473-2481.

Luoma, J.S., Stralin, P., Marklund, S.L., Hiltunen, T.P., Sarkioja, T. & Yla–Herttuala, S. (1998). Expression of Extracellular SOD and iNOS in Macrophages and Smooth Muscle Cells in Human and Rabbit Atherosclerotic Lesions: Colocalization With Epitopes Characteristic of Oxidized LDL and Peroxynitrite - Modifed Proteins. *Arterioscler Thromb Vasc Biol,* 18(2), 157-167.

Luz, M.R., Bertan, C.M., Binelli, M. & Lopes, M.D. (2006). In vitro PGF2alpha production by endometrium and corpus luteum explants from pregnant and non-pregnant diestrus bitches and placental explants from pregnant bitches. *Theriogenology*, 66, 1442-7.

Mann, T. & Keilin, D. (1938). Haemocuprein and Hepatocuprein, Copper-Protein Compounds of Blood and Liver in Mammals. *Proc Roy Soc Lond B,* 126, 303-315.

McCord, J.M. (2002). Superoxide dismutase in age and disease: an overview. *Meth Enzymol,* 349, 331-341.

McCord, J.M. & Fridovich, I. (1969). Superoxide dismutase. An enzymatic function for erythrocuprein (hemocuprein). *J Biol Chem,* 244, 6049-6055.

Makker, A., Bansode, F.W., Srivastava, V.M. L., & Singh, M.M. (2006). Antioxidant defense system during endometrial receptivity in the guinea pig: effect of ormeloxifene, a selective estrogen receptor modulator. *J Endocrinol,* 188(1), 121-134.

Marklund, S.L. (1992). Regulation by cytokines of extracellular superoxide dismutase and other superoxide dismutase isoenzymes in fibroblasts. *J Biol Chem,* 267, 6696–6701.

Marklund, S.L, Holme, E. & Hellner, L. (1982). Superoxide dismutase in extracellular fluids. *Clin Chim Acta,* 126, 41-51.

Marikovsky, M., Nevo, N., Vadai, E. & Harris-Cerruti, C. (2002). Cu/Zn superoxide dismutase plays a role in angiogenesis. *Int J Cancer*, 97(1), 34-41.

Marikovsky, M., Ziv, V., Nevo, N., Harris-Cerruti, C. & Mahler, O. (2003). Cu/Zn superoxide dismutase plays important role in immune response. *J Immunol,* 170(6), 2993-3001.

Massaad, C.A., Washington, T.M., Pautler, R.G. & Klann, E. (2009). Overexpression of SOD-2 reduces hippocampal superoxide and prevents memory deficits in a mouse model of Alzheimer's disease. *Proc Natl Acad Sci USA,* 106(32), 13576-13581.

McCord, J.M. & Fridovich I. (1969). Superoxide dismutase. An enzymic function for erythrocuprein (hemocuprein). *J Biol Chem,* 244, 6049-6055.

Meier, B., Scherk, C., Schmidt, M. & Parak, F. (1998). pH-dependent inhibition by azide and fluoride of the iron superoxide dismutase from *Propionibacterium shermanii*. *Biochem J,* 331, 403-407.

Melov, S., Schneider J.A., Day, B.J., Hinerfeld, D., Coskun, P., Mirra, S.S., Crapo, J.D. & Wallace, D.C. (1998). A novel neurological phenotype in mice lacking mitochondrial manganese superoxide dismutase. *Nat Genet,* 18(2), 159-163.

Melov, S., Ravenscroft, J, Malik, S., Gill, M.S., Walker, D.W., Clayton, P.E., Wallace, D.C., Malfroy, B., Doctrow, S.R. & Lithgow, G.J. (2000). Extension of life-span with superoxide dismutase/catalase mimetics. *Science,* 289(5484), 1567-1569.

Miao, L. & St Clair, D.K. (2009). Regulation of superoxide dismutase genes: implications in disease. *Free Radic Biol Med,* 47, 344-356.

Mollsten, A., Marklund, S.L., Wessman, M., Svensson, M., Forsblom, C., Parkkonen, M., Brismar, K., Groop, P.H. & Dahlquist, G. (2007). A functional polymorphism in the manganese superoxide dismutase gene and diabetic nephropathy. *Diabetes,* 56, 265-269.

Muller, F.L., Song, W., Liu, Y., Chaudhuri, A., Pieke-Dahl, S., Strong, R., Huang, T.T., Epstein, C.J., Roberts, L.J., Csete, M., Faulkner, J.A. & Van Remmen, H. (2006). Absence of CuZn superoxide dismutase leads to elevated oxidative stress and acceleration of age dependent skeletal muscle atrophy. *Free Radic Biol Med,* 40, 1993-2004.

Murakami, K., Inagaki, J., Saito, M., Ikeda, Y., Tsuda, C., Noda, Y. Kawakami, S., Shirasawa, T. & Shimizu, T. (2009). Skin atrophy in cytoplasmic SOD-deficient mice and its complete recovery using a vitamin C derivative. *Biochem Biophys Res Commun,* 382, 457-461.

Murakami, K., Murata, N., Noda, Y., Tahara, S., Kaneko, T., Kinoshita, N., Hatsuta, H., Murayama, S., Barnham, K.J., Irie, K., Shirasawa, T. & Shimizu, T. (2011). SOD1 (copper/zinc superoxide dismutase) deficiency drives amyloid beta protein oligomerization and memory loss in mouse model of Alzheimer disease. *J Biol Chem,* 286(52), 44557-44568.

Narimoto, K., Noda, Y. & Shiotani, M. (1990). Immunohistochemical assessment of superoxide-dismutase expression in the human endometrium throughout the menstrual cycle. *Acta Histochem Cytochemica,* 23, 487-498.

Nassif, J., Abbasi, S.A., Nassar, A., Abu-Musa, A. & Eid, A.A. (2016). The role of NADPH-derived reactive oxygen species production in the pathogenesis of endometriosis: a novel mechanistic approach. *J Biol Regul Homeost Agents,* 30(1), 31-40.

Nishikawa, M., Nagatomi, H., Nishijima, M., Ohira, G., Chang, B.J., Sato, E. & Inoue, M. (2001). Targeting superoxide dismutase to renal proximal tubule cells inhibits nephrotoxicity of cisplatin and increases the survival of cancer-bearing mice. *Cancer Lett,* 171, 133-138.

Nivsarkar, M., Patel, M., Padh, H., Bapu, C. & Shrivastava, N. (2005). Blastocyst implantation failure in mice due to "non-receptive endometrium": endometrial alterations by *Hibiscus rosa-sinensis* leaf extract. *Contraception*, 71(3), 227-230.

Noor, R., Mital, S. & Iqbal, J. (2002). Superoxide dismutase – applications and relevance to human diseases *Med Sci Monit*, 8(9), RA210-215.

Nonogaki, T., Noda, Y., Narimoto, K., Umaoka, Y & Mori, T. (1992). Effects of superoxide dismutase on mouse in vitro fertilization and embryo culture system. *J Assist Reprod Genet,* 9(3), 274-280.

Obal, D., Dai, S., Keith, R., Dimova, N., Kingery, J., Zheng, Y.T., Zweier, J., Velayutham, M., Prabhu, S.D., Li, Q., Conklin, D., Yang, D., Bhatnagar, A., Bolli, R. & Rokosh, G. (2012). Cardiomyocyte-restricted overexpression of EC-SOD increases NO• bioavailability and reduces infarct size after ischemia/reperfusion. *Basic Res Cardiol*, 107, 305.

Okada, H., Tsuzuki, T., Murata, H., Kasamatsu, A., Yoshimura, T. & Kanzaki, H. (2016). Regulation of angiogenesis in the human endometrium. In *Uterine Endometrial Function*. Kanzaki, H. (ed). Springer Japan. 83-103.

Orr, W.C. & Sohal, R.S. (1994). Extension of life-span by overexpression of superoxide dismutase and catalase in *Drosophila melanogaster*. *Science,* 263(5150), 1128-1130.

Page, M.M., Richardson, J., Wiens, B.E., Tiedtke, E., Peters, C.W., Faure, P.A., Burness, G. & Stuart, J.A. (2010). Antioxidant enzyme activities are not broadly correlated with longevity in 14 vertebrate endotherm species. *Age (Dordr.),* 32(2), 255-270.

Papalou, O., Victor, V.M. & Diamanti-Kandarakis, E. (2016). Oxidative stress in polycystic ovary syndrome. *Curr Pharm Des,* 22(18), 2709-2722.

Park, C.K., Jung, J.H., Moon, M.J., Kim, Y.Y., Kim, J.H., Park, S.H., Kim, C.Y., Paek, S.H., Kim, D.G., Jung, H.W. & Cho, B.K. (2009). Tissue expression of manganese superoxide dismutase is a candidate prognostic marker for glioblastoma. *Oncology,* 77(3–4), 178-181.

Pauling, L. (1979). The discovery of the superoxide radical. *Trends Biochem Sci*, 4, N270-N271.

Payan-Carreira, R., Santos, C., Miranda, S., Pereira, R.M. L.N., Santos, D. & Pires, M.D. A. (2014). Temporal changes in neutral endopeptidase/CD10 immunoexpression in the cyclic and early pregnant canine endometrium. *Theriogenology*, 82(6), 815-826.

Pejić, S., Todorović, A., Stojiljković, V., Kasapović, J. & Pajović, S.B. (2009). Antioxidant enzymes and lipid peroxidation in endometrium of patients with polyps, myoma, hyperplasia and adenocarcinoma. *Reprod Biol Endocrinol*, 7(1), 149.

Pereira, R.D., De Long, N.E., Wang, R.C., Yazdi, F.T., Holloway, A.C. & Raha, S. (2015). Angiogenesis in the placenta: the role of reactive oxygen species signaling. *Biomed Res Intl*, 2015.

Perumal, P. (2014). Effect of superoxide dismutase on semen parameters and antioxidant enzyme activities of liquid stored (5°C) Mithun (*Bos frontalis*) semen. *J Animals,* 2014, 9.

Perry, J.J., Shin, D.S. & Tainer, J.A. (2010). Amyotrophic lateral sclerosis. *Adv Exp Med Biol,* 685, 9–20.

Pertynska-Marczewska, M., Diamanti-Kandarakis, E., Zhang, J. & Merhi, Z. (2015). Advanced glycation end products: a link between metabolic and endothelial dysfunction in polycystic ovary syndrome? *Metabolism,* 64(11), 1564-1573.

Ponnampalam, A.P., Weston, G.C., Trajstman, A.C., Susil, B. & Rogers, P.A. (2004). Molecular classification of human endometrial cycle stages by transcriptional profiling. *Mol Hum Reprod,* 10, 879-893.

Prieto, L., Quesada, J.F., Cambero, O., Pacheco, A., Pellicer, A., Codoceo, R. & Garcia-Velasco, J.A. (2012). Analysis of follicular fluid and serum markers of oxidative stress in women with infertility related to endometriosis. *Fertil Steril,* 98, 126-130.

Pollak, R., Andrisevic, J.H., Maddux, M.S., Gruber, S.A. & Paller, M.S. (1993). A randomized double-blind trial of the use of human recombinant SOD in renal transplantation. *Transplantation,* 55, 57-60.

Rahiminejad, M.E., Moaddab, A., Ganji, M., Eskandari, N., Yepez, M., Rabiee, S., Wise M, Ruano, R. & Ranjbar, A. (2016). Oxidative stress biomarkers in endometrial secretions: A comparison between successful and unsuccessful in vitro fertilization cycles. *J Reprod Immunol,* 116, 70-75.

Ramos, R.S., Oliveira, M.L., Izaguirry, A.P., Vargas, L.M., Soares, M.B., Mesquita, F.S., Santos, F.W. & Binelli, M. (2015). The periovulatory endocrine milieu affects the uterine redox environment in beef cows. *Reprod Biol Endocrinol,* 13, 39.

Richters, L., Lange, N., Renner, R., Treiber, N., Ghanem, A., Tiemann, K., Scharffetter-Kochanek, K., Bloch, W. & Brixius, K. (2011). Exercise-induced adaptations of cardiac redox homeostasis and remodeling in heterozygous SOD2-knockout mice. *J Appl Physiol*, 111, 1431.

Rizk, B., Badr, M. & Talerico, C. (2013). Oxidative stress and the endometrium. In *Studies on Women's Health.* Agarwal, A., Aziz, N. and Rizk, B. (eds). Humana Press. Totowa, NJ, USA. 61-74.

Rodriguez-Iturbe, B., Sepassi, L., Quiroz, Y., Ni, Z., Wallace, D.C. & Vaziri, N.D. (2007). Association of mitochondrial SOD deficiency with salt-sensitive hypertension and accelerated renal senescence. *J Appl Physiol,* 102, 55-260.

Rosen, D.R., Siddique, T., Patterson, D., Figlewicz, D.A., Sapp, P., Hentati, A., Donaldson, D., Goto, J., O'Regan, J.P., Deng, H.X. et al. (1993). Mutations in Cu/Zn superoxide dismutase gene are associated with familial amyotrophic lateral sclerosis. *Nature,* 362(6415), 59-62.

Salway, K.D., Page, M.M., Faure, P.A., Burness, G. & Stuart, J.A. (2010). Enhanced protein repair and recycling are not correlated with longevity in 15 vertebrate endotherm species. *Age (Dordr.),* 33(1), 33-47.

Sandoval, J.A., Hoelz, D.J., Woodruff, H.A., Powell, R.L., Jay, C.L., Grosfeld, J.L., Hickeyd, R.J. & Malkas, L.H. (2006). Novel peptides secreted from human neuroblastoma: useful clinical tools? *J Pediatr Surg,* 41(1), 245-251.

Santos, C., dos Anjos Pires, M., Santos, D. & Payan-Carreira, R. (2016). Distribution of superoxide dismutase 1 and glutathione peroxidase 1 in the cyclic canine endometrium. *Theriogenology*, 86(3), 738-748.

Seleem, A.K., El Refaeey, A.A., Shaalan, D., Sherbiny, Y. & Badawy, A. (2014). Superoxide dismutase in polycystic ovary syndrome patients undergoing intracytoplasmic sperm injection. *J Assist Reprod Genet,* 31(4), 499-504.

Serviddio, G., Loverro, G., Vicino, M., Prigigallo, F., Grattagliano, I., Altomare, E. & Vendemiale, G. (2002). Modulation of endometrial redox balance during the menstrual cycle: relation with sex hormones. *J Clin Endocrinol Metab,* 87(6), 2843-2848.

Shaman A., Premkumar B.J. & Agarwal A. (2013). Placental Vascular Morphogenesis and Oxidative Stress. In *Studies on Women's Health.* Agarwal A., Aziz N. and Rizk B. (eds). Humana Press, Totowa, NJ. USA. 95-113.

Sharma, J.B., Sharma, A., Bahadur, A., Vimala, N., Satyam, A. & Mittal, S. (2006). Oxidative stress markers and antioxidant levels in normal pregnancy and pre-eclampsia. *Int J Gynaecol Obstet*, 94(1), 23-27.

Sharma, R., Roychoudhury, S., Alsaad, R. & Bamajbuor, F. (2017). Negative Effects of Oxidative Stress (OS) on Reproductive System at Cellular Level. In *Oxidative Stress in Human Reproduction.* Springer International Publishing. 65-87.

Shen, X., Zheng, S., Metreveli, N.S. & Epstein, P.N. (2006). Protection of cardiac mitochondria by overexpression of MnSOD reduces diabetic cardiomyopathy. *Diabetes* 55(3), 798–805.

Sherwin, R., Catalano, R. & Sharkey, A. (2006). Large-scale gene expression studies of the endometrium: what have we learnt? *Reproduction,* 132, 1-10.

Shigetomi, H., Higashiura, Y., Kajihara, H. & Kobayashi, H. (2012). A potential link of oxidative stress and cell cycle regulation for development of endometriosis. *Gynecol Endocrinol,* 28(11), 897-902.

Sies, H., Berndt, C. & Jones, D.P. (2017). Oxidative stress. *Annu Rev Biochem,* 86, 715-748.

Smith, S.K. (2001). Regulation of angiogenesis in the endometrium. *Trends Endocrinol Metab*, 12(4), 147-151.

Sohal, R.S., Agarwal, A., Agarwal, S. & Orr, W.C. (1995). Simultaneous overexpression of copper- and zinc-containing superoxide dismutase and catalase retards age-related oxidative damage and increases metabolic potential in *Drosophila melanogaster*. *J Biol Chem,* 270(26), 15671-15674.

St. Clair, D.K., Oberley, T.D. & Ho, Y.S. (1991). Overproduction of human Mn-superoxide dismutase modulates paraquat-mediated toxicity in mammalian cells. *FEBS Lett,* 293, 199-203.

Sugino, N. (2007). The role of oxygen radical-mediated signaling pathways in endometrial function. *Placenta,* 28, S133-S136.

Sugino, N., Shimamura, K., TakiguchI, S., Tamura, H., Ono, M., Nakata, M., Nakamura, Y., Ogino, K., Uda, T. & Kato, H (1996). Changes in activity of superoxide dismutase in the human endometrium throughout the menstrual cycle and in early pregnancy. *Hum Reprod,* 11, 1073e8.

Sugino, N., Kashida, S., Takiguchi, S., Nakamura, Y. & Kato, H. (2000). Induction of superoxide dismutase by decidualization in human endometrial stromal cells. *Mol Hum Reprod,* 6, 178e84.

Sugino, N., Karube-Harada, A., Kashida, S., Takiguchi, S. & Kato, H. (2001). Reactive oxygen species stimulate prostaglandin F2 alpha production in human endometrial stromal cells in vitro. *Hum Reprod,* 16, 1797-1801.

Sugino, N., Karube-Harada, A., Taketani, T., Sakata, A. & Nakamura, Y. (2004). Withdrawal of ovarian steroids stimulates prostaglandin F2alpha production through nuclear factor-kappaB activation via oxygen radicals in human endometrial stromal cells: potential relevance to menstruation. *J Reprod Develop*, 5, 215-25.

Suresh, S. & Vijayakumar, T. (2015). Correlations of insulin resistance and serum testosterone levels with LH: FSH ratio and oxidative stress in women with functional ovarian hyperandrogenism. *Indian J Clin Biochem,* 30(3), 345-350.

Taylor, A. (1993). Cataract: relationship between nutrition and oxidation. *J Am Coll Nutr*, 12, 138.

Teoh, M.L., Fitzgerald, M.P., Oberley, L.W. & Domann, F.E. (2009). Overexpression of extracellular superoxide dismutase attenuates heparanase expression and inhibits breast carcinoma cell growth and invasion. *Cancer Res,* 69, 6355-6363.

Thannickal, V.J. & Fanburg, B.L. (2000). Reactive oxygen species and cell signaling. *Am J Physiol Lung Cell Mol Physiol,* 279, L1005-L1028.

Torry, D.S., Leavenworth, J., Chang, M., Maheshwari, V., Groesch, K., Ball, E.R. & Torry, R.J. (2007). Angiogenesis in implantation. *J Assist Reprod Genet*, 24(7), 303-315.

Tsunoda, S., Kawano, N., Miyado, K., Kimura, N. & Fujii, J. (2012). Impaired fertilizing ability of superoxide dismutase 1-deficient mouse sperm during in vitro fertilization. *Biol Reprod*, 87(5), 121.

Uchiyama, S., Shimizu, T. & Shirasawa, T. (2006). CuZn-SOD deficiency causes ApoB degradation and induces hepatic lipid accumulation by impaired lipoprotein secretion in mice. *J Biol Chem,* 281, 31713-31719.

Valko, M., Leibfritz, D., Moncol, J., Cronin, M., Mazur, M. & Telser, J. (2007). Free radicals and antioxidants in normal physiological functions and human disease. *Int J Biochem Cell Biol,* 39, 44-84.

Valentine, J.S., Doucette, P.A. & Zittin Potter, S. (2005). Copper–zinc superoxide dismutase and amyotrophic lateral sclerosis. *Annu Rev Biochem,* 74, 563–593.

Van Remmen, H., Ikeno, Y., Hamilto, M., Pahlavani, M., Wolf, N., Thorpe, S.R., Alderson, N.L., Baynes, J.W., Epstein, C.J., Huang, T.T., Nelson, J., Strong, R. & Richardson, A. (2003). Life-long reduction in MnSOD activity results in increased DNA damage and higher incidence of cancer but does not accelerate aging. *Physiol Genomics,* 16(1), 29-37.

Victor, V.M., Rovira-Llopis, S., Banuls, C., Diaz-Morales, N., Martinez de Maranon, A., Rios-Navarro, C., Alvarez, A., Gomez, M., Rocha, M. & Hernandez-Mijares, A. (2016). Insulin resistance in PCOS patients enhances oxidative stress and leukocyte adhesion: role of myeloperoxidase. *PLoS One,* 11(3), e0151960.

Vincent, A.M., Russell, J.W., Sullivan, K.A., Backus, C., Hayes, J.M., McLean, L.L. & Feldman, E.L. (2007). SOD2 protects neurons from injury in cell culture and animal models of diabetic neuropathy. *Exp Neurol,* 208, 216-227.

Warner, H.R. (1994). Superoxide dismutase, aging and degenerative disease. *Free Rad Biol Med*, 17(3), 249-258.

Warner, A., Bencosme, A., Healy, D. & Verme, C. (1995). Prognostic role of antioxidant enzymes in sepsis: preliminary assessment, *Clin Chem,* 41(61), 867–871.

Wheeler, M.D., Nakagami, M., Bradford, B.U., Uesugi, T., Mason, R.P., Connor, H.D., Dikalova, A., Kadiiska, M. & Thurman, R.G. (2001). Overexpression of manganese superoxide dismutase prevents alcohol-induced liver injury in the rat. *J Biol Chem,* 276(39), 36664-36672.

Wu, B., Chen, X., He, B., Liu, S., Li, Y., Wang, Q., Gao, H., Wang, S., Liu, J., Zhang, S. & Xu, X. (2014). ROS are critical for endometrial breakdown via NF-κB–COX-2 signaling in a female mouse menstrual-like model. *Endocrinology*, 155(9), 3638-3648.

Wuerges, J., Lee, J.W., Yim, Y.I., Yim, H.S., Kang, S.O. & Djinovic Carugo, K. (2004). Crystal structure of nickel-containing superoxide dismutase reveals another type of active site. *Proc Natl Acad Sci. USA,* 101, 8569-8574.

Yamakura, F. & Kawasak, H. (2010). Post-translational modifications of superoxide dismutase. *Biochim Biophys Acta,* 804, 318-325.

Yao, H., Arunachalam, G., Hwang, J.W., Chung, S., Sundar, I.K., Kinnula, V.L., Crapo, J.D. & Rahman, I. (2010). Extracellular superoxide dismutase protects against pulmonary emphysema by attenuating oxidative fragmentation of ECM. *Proc Natl Acad Sci USA,* 107(35), 15571-15576.

Yen, H.C., Oberley, T.D., Vichitbandha, S., Ho, Y.S. & St Clair, D.K. (1996). The protective role of manganese superoxide dismutase against adriamycin-induced acute cardiac toxicity in transgenic mice. *J Clin Invest,* 98(5), 1253-1260.

Yin, M., Wheeler, M.D., Connor, H.D., Zhong, Z., Bunzendahl, H., Dikalova, A., Samulski, R.J., Schoonhoven, R., Mason, R.P., Swenberg, J.A. & Thurman, R.G. (2001). Cu/Zn-superoxide dismutase gene attenuates ischemia-reperfusion injury in the rat kidney. *J Am Soc Nephrol,* 12(12), 2691-2700.

Yost, F.J. & Fridovich, I. (1973). An iron-containing superoxide dismutase from *Escherichia coli*. *J Biol Chem,* 248, 4905-4908.

Zelko, I.N., Mariani, T.J. & Folz, R.J. (2002). Superoxide dismutase multigene family: a comparison of the CuZn - SOD (SOD1), Mn - SOD (SOD2), and EC - SOD (SOD3) gene structures, evolution, and expression. *Free Radic Biol Med,* 33, 337-349.

Zhang, J., Wang, X., Vikash, V., Ye, Q., Wu, D., Liu, Y. & Dong, W. (2016). ROS and ROS-Mediated Cellular Signaling. *Oxid Med Cell Longev*, 2016, 1-18.

Zuo, T., Zhu, M. & Xu, W. (2016). Roles of oxidative stress in polycystic ovary syndrome and cancers. *Oxid Med Cell Longev,* 2016, 8589318.

In: Superoxide Dismutase
Editor: Sergei Magliozzi
ISBN: 978-1-53613-080-5

Chapter 2

PLANT SUPEROXIDE DISMUTASES: IMPORTANT PLAYERS IN ABIOTIC STRESS TOLERANCE

Ivana Momčilović** and Danijel Pantelić
Department of Plant Physiology, Institute for Biological Research
"Siniša Stanković", University of Belgrade, Belgrade, Serbia

ABSTRACT

Abiotic stresses affect plant growth and development, and significantly reduce crop productivity by causing the alteration in cell biochemistry. Common for all abiotic stresses is that they cause an imbalance between production and scavenging of reactive oxygen species (ROS) in cells, resulting in elevation of ROS concentrations to cytotoxic level. Superoxide dismutases (SODs) are enzymes that play an important role of primary ROS scavengers by catalyzing the conversion of superoxide anion radical to hydrogen peroxide and oxygen. Most plant species contain numerous SOD isoforms which are classified according to the active site metal into three major types: FeSOD, MnSOD and Cu/ZnSOD. All plant SODs are nuclear-encoded, synthesized on

* Corresponding Author Email: ivana.momcilovic@ibiss.bg.ac.rs.

cytosolic ribosomes and transported into various cell compartments, such as cytosol, plastids, mitochondria, peroxisomes and cell wall. Expression of SODs genes might be modulated by developmental and environmental cues, including nutrition and abiotic stresses. So far, enhanced expression and/or activity of SODs have been reported in response to various abiotic stressors, such as water deficit, high salinity, heavy metals, hypoxia, high and low temperatures. This chapter summarizes the current knowledge of plant SODs, their abiotic-stress modulated expression and activity, and analyses results on genetic engineering of plant SODs. The significance of superoxide dismutases in the crop improvement for stress tolerance is also discussed.

INTRODUCTION

Abiotic stresses affect plant growth and development, and significantly reduce crop productivity. Common for all abiotic stresses is that they promote oxidative stress, the physiological condition where imbalance between production and scavenging of reactive oxygen species (ROS) results in elevation of ROS concentrations in the cell to cytotoxic level. ROS, such as superoxide anion radical ($O_2^{\bullet -}$), hydrogen peroxide (H_2O_2), hydroxyl radical ($OH^{\bullet}$), alkoxyl radical ($RO^{\bullet}$) and peroxyl radical ($ROO^{\bullet}$), can damage cell lipids, proteins and DNA (biomolecules), leading ultimately to cellular metabolism arrest. Plant response to elevated ROS production is the induction of antioxidant defense system which includes non-enzymatic radical scavengers, such as ascorbic acid, glutathione, phenolic compounds, α-tocopherol and carotenoids, and antioxidant enzymes, such as superoxide dismutase, catalase, glutathione reductase, monodehydroascorbate reductase, dehydroascorbate reductase, and guaiacol, glutathione and ascorbate peroxidases.

Superoxide dismutase (SOD, EC 1.15.1.1) is a metalloenzyme that plays an important role of primary ROS scavenger in the antioxidant defense system of plant cells. SOD catalyzes conversion of superoxide anion radical ($O_2^{\bullet -}$) to hydrogen peroxide and oxygen in the redox reaction that involves active-site metal ion. The half-reactions that occur are given below:

$$O_2^{\bullet-} + M^{(n+1)+}\text{-SOD} \rightarrow O_2 + M^{n+}\text{-SOD}$$

$$O_2^{\bullet-} + M^{n+}\text{-SOD} + 2H^+ \rightarrow H_2O_2 + M^{(n+1)+}\text{-SOD}$$

where M represents SOD active site metal ion: Cu (n = 1), Mn (n = 2), Fe (n = 2) or Ni (n = 2).

The overall dismutation reaction can be presented as the following:

$$2O_2^{\bullet-} + 2H^+ \rightarrow H_2O_2 + O_2$$

The outcome of this reaction is the production of H_2O_2, ROS that can be efficiently detoxified, and prevention of $O_2^{\bullet-}$ participation in the metal-catalyzed Haber-Weiss–type reactions which yield highly reactive and damaging $OH^{\bullet}$. The H_2O_2 is converted to water by the action of two other antioxidant enzymes, catalase (CAT) and glutathione peroxidase (GPX).

SOD is considered a key enzyme in the regulation of intracellular ROS levels and the maintenance of plant cell redox homeostasis under oxidative stress. Increase in SOD expression and activity has been detected in a large number of plant species in response to various abiotic stressors, such as water deficit, high salinity, heavy metals, hypoxia, high and low temperatures. Improved tolerances of some SOD-overexpressing transgenic plants suggest that enhanced expression of SOD can protect plants against various abiotic stresses.

This chapter summarizes the current knowledge of plant superoxide dismutases, with emphasis on molecular mechanisms of gene expression and modulation of SOD expression by abiotic stresses. Results on genetic engineering of plant SODs were analyzed and significance of SOD antioxidant enzymes in the plant improvement for stress tolerance was also discussed.

Types, Structure and Localization of Plant Superoxide Dismutases

Based on the active site metal, the three types of SOD have been identified in plants: FeSOD (iron cofactor), MnSOD (manganese cofactor) and Cu/ZnSOD (copper redox active metal; zinc maintains structural stability). NiSOD is the fourth group of SOD, detected in cyanobacteria, marine gammaproteobacteria and unicellular eukaryotes (Dupont et al., 2008), but not in vascular plants. SOD overall dismutation reaction involves the binding of two $O_2^{\bullet-}$ and two redox-active metal ions for catalysis, thus SOD functions as dimer with a metal ion in each monomer (Abreu and Cabelli, 2010; Pilon et al., 2011). This was confirmed by structural analysis of cowpea FeSOD which revealed close proximity of monomer active sites to the dimer interface (18.01 Å between the two Fe atoms) and possible participation of certain amino acid residues of one monomer in the catalysis on the active site of the other (Muñoz et al., 2005). SODs are also present in the form of tetramers which probably function as a pair of dimers that undergo catalysis in an independent manner (Muñoz et al., 2005). Interestingly, monomeric Cu/ZnSODs were reported in several plant species, rice, aspen and weed *Chenopodium murale* (Kanematsu and Asada, 1989; Schinkel et al., 2001; Sundaram et al., 2009).

Plants contain numerous SOD isoforms (Momčilović et al., 2014) located in various cellular compartments. Beside cytosol, SODs are mainly located in organelles which are major ROS-producing sites in the cell, i.e., mitochondria, chloroplast and peroxisomes. SODs can also occur in nucleus and extracellular space.

Iron Superoxide Dismutase

FeSOD is probably the most ancient type of SOD since the soluble Fe (II) iron was the most abundant metal in the environment at the time when evolution of oxidizing photosynthesis started, as well as a rise in oxygen levels. The transition from reducing to the oxidizing environment made iron, at that time mostly in oxidized Fe (III) form, less bioavailable and probably caused a shift toward usage of Mn (III) as SOD active site metal (Alscher et al., 2002).

FeSOD is present in prokaryotes, primitive eukaryotes' mitochondria and cytosol, as well as plant plastids (Figure 1.). The plastid occurrence of this ancient SOD, as well as absence in animals and fungi, gave rise to the hypothesis that FeSOD gene originated from cyanobacterial endosymbiont and moved to the nuclear genome during the evolution of plant cell. Support for this hypothesis came from detection of several conserved regions in both plant and cyanobacterial FeSOD-gene sequences, which are absent in non-photosynthetic bacteria FeSOD genes (Bowler et al., 1994). However, recent findings indicate that FeSOD can also be cytosol-localized in plants. Atypical cytosol-localized FeSOD was found in *Vigna unguiculata* (Moran et al., 2003), *Lotus japonicus* (Rubio et al., 2007) and moss *Pogonatum inflexum* (Kanematsu et al., 2013). Besides, the cellular location of *Arabidopsis* FSD1 isoform is not fully resolved due to the absence of chloroplast transit sequence (Pilon et al., 2011).

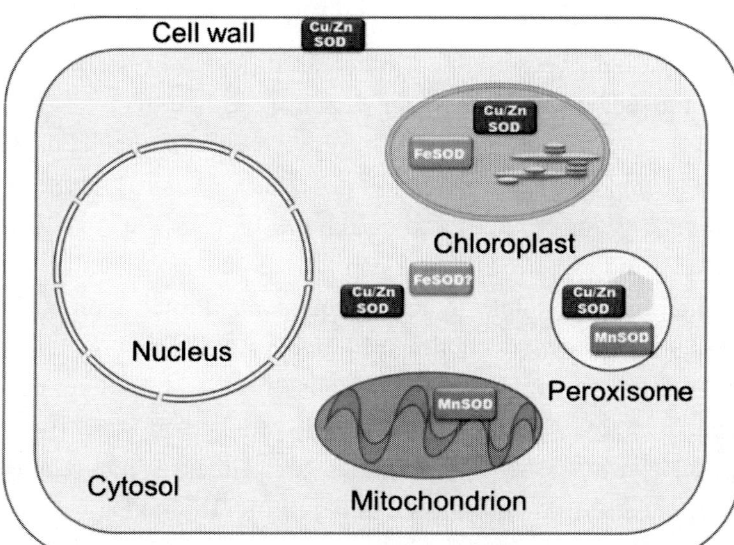

Figure 1. Localization of superoxide dismutases in plant cell.

Two distinct groups of FeSOD are present in plants. The first group comprises FeSODs that form homodimer from two identical ~18-27 kDa subunits with 1-2 gram atom of iron in the active center. This group was found in *Brassica campestris*, *Nuphar luteum*, *Ginkgo biloba*, *Vigna unguiculata*, *Pisum sativum* (reviewed by Alscher et al., 2002). The second FeSOD group, found in most vascular plants, comprises 80-90 kDa-homotetramers of four identical subunits and 2-4 gram atoms of iron in the active center. Recent findings indicate that FeSODs could also have heterodimeric structure. Heterocomplexes of FeSODs, composed of *A. thaliana* recombinant FSD2 and FSD3 protein subunits, were detected by the bimolecular fluorescence complementation (BiFC) in plastids of transiently transformed tobacco leaves (Myouga et al., 2008). FeSODs protect chloroplasts against oxidative stress, and probably play an important role in chloroplast development in *Arabidopsis* (Myouga et al., 2008).

Manganese Superoxide Dismutase

MnSOD most likely evolved from FeSOD as the level of bioavailable Fe (II) in oxidizing environment decreased and Mn (III) became a substitute. Indeed, SODs that can function either with Fe or Mn, Fe/MnSOD, are found in relatively primitive anaerobic bacteria, such as *Propionibacterium shermanii* (Meier et al., 1994), *Bacteroides fragilis* and *Porphyromonas gingivalis* (Amano et al., 1990). These SODs are termed cambialistic. 'Modern' FeSOD and MnSOD could have evolved from a common ancestral SOD able to use either metal. Protein components of these metalloenzymes show significant homology in amino acid sequence, have similar fold and use conserved residues for metal binding, but use a single type of active site metal in the fully functional state (Fridovich, 1986). Similarly to FeSODs, MnSODs are either a homodimeric or homotetrameric enzymes with the subunit size of ~18-20 kDa.

MnSODs are found in prokaryotes and eukaryotic organelles, mitochondria and peroxisomes (Figure 1). In cyanobacteria and photosynthetic protozoa, MnSODs are found in the thylakoid membrane (Okada et al., 1979; Kanematsu and Asada, 1979). In the eukaryotic alga *Chlamydomonas rheinhardtii*, MnSOD occurs in the chloroplast (Allen et al., 2007), but it is not found in the plastids of higher plants. This suggests that MnSOD gene of cyanobacterial endosymbiont was probably lost during evolution in plants, while MnSOD gene from mitochondrial endosymbiont was retained. Indeed, MnSOD is frequently referred to as mitochondrial SOD in plants and it is detected in mitochondrial matrix of the number of species, including *Zea mays*, *Nicotiana tabacum*, *Vigna mungo*, *Dianthus caryophyllus*, *Spinacia oleracea* (reviewed in Alscher et al., 2002). Peroxisomal forms of MnSOD were also found in several plants species, including *Pisum sativum*, *Dianthus caryophyllus*, watermelon, castor bean and cucumber (del Río et al., 2002). Besides the important role of MnSOD in processing $O_2^{\bullet-}$ in mitochondria and peroxisomes during plant growth and development, MnSOD abundance and enzymatic activity significantly affect leaf senescence in maize and rice (Prochazkova et al., 2001; Wang et al., 2016a).

Copper-Zinc Superoxide Dismutase

Cu/ZnSOD is the most abundant type of plant SODs. Cu/ZnSOD enzyme can be considered like evolutionary novelty compared to FeSOD and MnSOD. In oxygen replenished atmosphere, insoluble Cu (I) was converted into soluble Cu (II) which started to be used as SOD metal cofactor (Alscher et al., 2002). Cu/ZnSOD protein appears to have evolved independently from mutually related FeSOD and MnSOD. For instance, comparison of *A. thaliana* Cu/ZnSODs shows 51% identity and 80% similarity in deduced amino acid sequence, while the comparison between *A. thaliana* Cu/ZnSOD (CSD1) and FeSOD (FSD1) reveals only 14% identity (Figure 2.). Cu/ZnSOD forms distinctive Greek key β-barrel, while FeSOD and MnSOD both contain α/β fold (Figure 3.).

Cu/ZnSODs are found in different plant cell compartments, including cytosol, plastids, mitochondria, peroxisomes, as well as extracellular space (Figure 1). Plant Cu/ZnSODs can be homodimeric (cytosolic) or homotetrameric (chloroplastic and extracellular), build from ~15-17 kDa subunits. Based on amino acid sequence, plant Cu/ZnSODs are divided into cytosolic and chloroplastic forms. Comparison between chloroplastic and cytosolic isoforms shows 68% similarity in deduced amino acid sequence. Similarity among cytoplasmic isoforms is 80-90%, while chloroplastic are approximately 90% similar in sequence (Alscher et al., 2002). An exception is the cytosolic Cu/ZnSOD of the liverwort *Marchantia paleacea* (MSODCc) which shows a very high homology with chloroplastic Cu/ZnSODs (Tanaka et al., 1996). Considering that the genus *Marchantia* is a very old taxon, it was proposed that MSODCc could be the surviving SOD that is closest to the SOD common ancestor of plant cytoplasmic and chloroplast Cu/ZnSODs (Fink and Scandalios, 2002).

Chloroplastic Cu/ZnSOD is a soluble enzyme, localized in chloroplast stroma (Kanematsu and Asada, 1990). Immunohistochemical studies on *Spinacia oleracea* showed that chloroplastic Cu/ZnSOD is localized mainly on the stromal face of thylakoid membranes, in the proximity of photosystem I (PSI) (Ogawa et al., 1995). Cytosolic isoform, on the other hand, was detected in the cytosol, nucleus and extracellular space (apoplast) (Ogawa et al., 1996). Extracellular Cu/ZnSODs, found in pine (Karpinska et al., 2001) and aspen (Schinkel et al., 2001), are probably involved in lignification of secondary walls. Findings regarding cotton fiber cell wall biosynthesis suggest that extracellular Cu/ZnSOD and peroxidases regulate extracellular hydrogen peroxide levels and, consequently, the processes of wall loosening and wall tightening (Kim and Triplett, 2008). An isoform of Cu/ZnSOD was also detected in peroxisomes of watermelon cells (Bueno et al., 1995). In *Arabidopsis*, CSD3 isoform of Cu/ZnSOD possesses C-terminal Ala-Lys-Leu motif which, most likely, represents a peroxisomal targeting sequence. Also, the CSD3 encoding gene is abundantly expressed in flowers, which are enriched in peroxisomes (Chu et al., 2005).

```
CSD1           ............................................................
CSD2           ....MAATNTILAFSSPSRLLIPPSSNPSTLRSSFRGVSLNNNNLHRLQSVSFAVKAPSKAL
CSD3           ...........................................................M
FSD1           ............................................................
FSD2           ............................................................
FSD3           MSSCVVTTSCFYTISDSSIR.LKSPKLLNLSNQQRRRSLRSRGGLKVEAYYGLKTPPYPLDA
MSD1           ................MAIRCVASRKTLAGLKETSSRLLRIRG....IQTFTLPDLPYDYGA
consensus>50   ............................................................

                    1        10        20        30        40        50
CSD1           .....MAKGVAVLNSSEGVTGTIFFTQEGDGVTTVSGTVSGLKPGLHGFHVHALGDTTNGCM
CSD2           TVVSAAKKAVAVLKGTSDVEGVVTLTQDDSGPTTVNVRITGLTPGPHGFHLHEFGDTTNGCI
CSD3           EAPRGNLRAVALIAGDNNVRGCLQFVQDISGTTHVTGKISGLSPGFHGFHIHSFGDTTNGCI
FSD1           ....MSKQTLEFHWGKHHRAYVDNLKKQVLGTE.LEGKPLEHIIHSTYNNGDLLPAFNNAAQ
FSD2           ....MSRETLDYHWGKHHKTYVENLNKQILGTD.LDALSLEEVVLLSYNKGNMLPAFNNAAQ
FSD3           LEPYMSRRTLEVHWGKHHRGYVDNLNKQLGKDDRLYGYTMEELIKATYNNGNPLPEFNNAAQ
MSD1           LEPAISGEIMQIHHQKHHQAYVTNYNNALEQLD....QAVNKGDASTVVKLQSAIKFNGGGH
consensus>50   .....s...levh.gkhh..yv.nl.qq..g.d...g...e.........n.l.dfnng..

                 60        70        80        90       100       110
CSD1           STGPHFNPDGKTHGAPEDANRHAGDLGNITVGDDGTATFTITDCQIPLTGPNSIVGRAVVVH
CSD2           STGPHFNPNNMTHGAPEDECRHAGDLGNINANADGVAETTIVDNQIPLTGPNSVVGRAFVVH
CSD3           STGPHFNPLNRVHGPPNEEERHAGDLGNILAGSNGVAEILIKDKHIPLSGQYSILGRAVVVH
FSD1           AWNHEFFWESMKP...GGGGKPSGELLALLERDFTSYEKFYEEFNAAAATQFGAGWAWLAYS
FSD2           AWNHEFFWESIQP...GGGGKPTGELLRLIERDFGSFEEFLERFKSAAASNFGSGWTWLAYK
FSD3           VYNHDFFWESMQP...GGGDTPQKGVLEQIDKDFGSFTNFREKFTNAALTQFGSGWVWLVLK
MSD1           V.NHSIFWKNLAPSSEGGGEPPKGSLGSAIDAHFGSLEGLVKKMSAEGAAVQGSGWVWLGLD
consensus>50   ..nh.ffwe.m.p...ggge.p.gdlgn..e.dfgs.e..i.d.......qfg.gw.wlv..

              120              130       140        150
CSD1           ADPDDL.........GKGGHELSLATGNAGGRVACGIIGLQG....................
CSD2           ELKDDL.........GKGGHELSLTTGNAGGRLACGVIGLTPL...................
CSD3           ADPDDL.........GKGGHKLSKSTGNAGSRVGCGIIGLQSSADAKL..............
FSD1           ...............NEKLKVVKTPNAVNPLVL...GSFPLLTIDVWEHAYYLDFQNRRP
FSD2           ANRLDVANAVNPLPKEEDKKLVIVKTPNAVNPLVW...DYSPLLTIDTWEHAYYLDFENRRA
FSD3           ...............REERRLEVVKTSNAINPLVW...DDIPIICVDVWEHSYYLDYKNDRA
MSD1           ...............KELKKLVVDTTANQ.DPLVTKGGSLVPLVGIDVWEHAYYLQYKNVRP
consensus>50   ....d..............l.v..T.Na.nplv....gl.pl..id.weh.yyldf.n.r.

CSD1           ............................................................
CSD2           ............................................................
CSD3           ............................................................
FSD1           DYIKTFMTNLVSWEAVSARLEAAKAASA................................
FSD2           EYINTFMEKLVSWETVSTRLESAIARAVQREQEGTETEDEENPDDEVPEVYLDSDIDVSEVD
FSD3           KYINTFLNHLVSWNAAMSRMARAEAFVNLGEPNIPIA.......................
MSD1           EYLKN.VWKVINWKYASEVYEKENN...................................
consensus>50   .y........v.w...............................................
```

Figure 2. Multiple sequence alignments of seven SODs in *Arabidopsis thaliana*. Deduced amino acid sequence alignment was performed with Multialin (Corpet, 1988) and viewed with ESPript (Robert and Gouet, 2014). Regions of identity (shaded) and similarity (squared) are indicated. CSD1 - *Arabidopsis thaliana* cytosolic Cu/ZnSOD (UniProtKB - P24704); CSD2 - *Arabidopsis thaliana* chloroplastic Cu/ZnSOD (UniProtKB - O78310); CSD3 – *Arabidopsis thaliana* peroxisomal Cu/ZnSOD (UniProtKB - Q9FK60); FSD1 - *Arabidopsis thaliana* FeSOD (UniProtKB - P21276); FSD2 - *Arabidopsis thaliana* chloroplastic FeSOD (UniProtKB - Q9LU64); FSD3 - *Arabidopsis thaliana* chloroplastic FeSOD (UniProtKB - Q9FMX0); MSD1 - *Arabidopsis thaliana* mitochondrial MnSOD (UniProtKB - O81235).

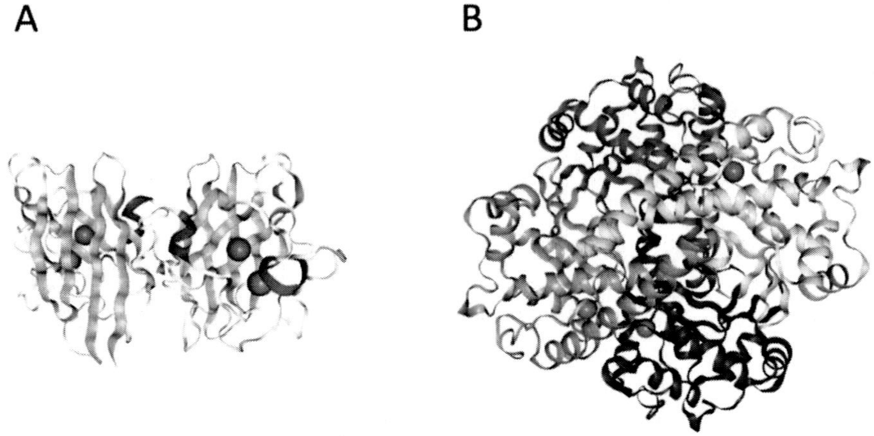

Figure 3. Comparison of Cu/ZnSOD and MnSOD ribbon structures. (A) *Solanum lycopersicum* chloroplastic Cu/ZnSOD is homodimeric protein with monomers that form flattened β-barrel (PDB entry 3PU7). The Cu and Zn ions are presented as spheres. (B) *Arabidopsis thaliana* MnSOD is homotetrameric with monomers that form α/β fold. The Mn ions are presented as spheres. (PDB entry 4C7U). The figure was prepared with NGL Viewer (Rose and Hildebrand, 2015) using Protein Data Bank (PDB).

Regulation of Plant SODs Expression

All three types of plant SODs, FeSOD, MnSOD and Cu/ZnSOD, are encoded by the nuclear genome, and organelle-specific forms are targeted to their appropriate compartment by transit peptides. The number of SOD-encoding genes notably varies among plants species. Four SOD genes (two *Cu/ZnSOD* and two *MnSOD*) were found in sunflower, seven (three *Cu/ZnSOD*, three *FeSOD* and one *MnSOD*) in the genome of *Arabidopsis*, while banana genome contains 12 SOD genes (six *Cu/ZnSOD*, two *FeSOD* and four *MnSOD*) (Kliebenstein et al., 1998; Fernández-Ocaña et al., 2011; Feng et al., 2015). Exon-intron organizations of Cu/ZnSOD and MnSOD genes are thought to be highly conserved in plants (Fink and Scandalios, 2002), with seven introns in genes encoding cytosolic Cu/ZnSODs, eight introns in genes encoding chloroplastic SODs and five in MnSOD genes. Recent findings, however, indicate significant variation in exon-intron structures of CuZnSODs in tomato (Feng et al., 2016), cucumber (Zhou et al., 2017) and upland cotton (Wang et al., 2017). Current characterizations of SOD gene families at the genome-wide level could help in determining full complements of SOD-encoding genes in various plant species and give insight into gene structure and possible molecular mechanisms of SOD genes regulation. So far, the genome-wide investigations of SODs were conducted in banana, tomato, upland cotton and cucumber (Feng et al., 2015; Feng et al., 2016; Wang et al., 2017; Zhou et al., 2017).

Various molecular mechanisms are involved in regulation of SODs expression at transcriptional, posttranscriptional, translational and posttranslational level. Expression of SODs genes shows temporal and special specificity and can be modulated by environmental cues, including metal availability and nutrition, biotic and abiotic stresses.

Molecular Mechanisms Involved in the Regulation of SOD Expression

Plant cells regulate the expression of their genes mainly by determining the rate at which they are transcribed into mRNAs. Transcriptional regulation usually includes complex networks of cooperating and interacting transcription factors and their co-regulators. Transcription factors (*trans*-acting factors) bind directly or through an intermediate protein to the gene at a specific conserved motif in sequence, called a *cis*-regulatory element. Characterization of SOD gene families at the genome-wide level revealed the existence of numerous *cis*-elements responsible for regulation of gene transcription in SOD genes (Lin and Lai, 2013; Feng et al., 2015; Feng et al., 2016; Wang et al., 2017; Zhou et al., 2017). SOD gene promoters might possess hormone-responsive elements, light-responsive elements, stress-responsive elements and several *cis*-elements related to developmental processes, besides conserved TATA and CAAT boxes. However, number of *cis*-elements of the above-mentioned groups may vary among SOD genes, and particular SOD genes may not possess elements belonging to the certain group. This indicates differences in gene tuning and differential expression in response to various factors. In longan, hormone-responsive elements were not found in gene encoding MnSOD, *DlMSD* (Lin and Lai, 2013). Conversely, promoter of cytosolic Cu/ZnSOD gene, *DlCSD1a*, contained the greatest number of hormone-responsive elements, including eight *cis*-acting regulatory elements associated with response to auxin, gibberellins, salicylic acid, methyl jasmonate and ethylene, suggesting that *DlCSD1a* plays a major role in hormone signal transduction during longan somatic embryogenesis (Lin and Lai, 2013). Besides, the *DlCSD1a* was the only longan SOD gene which contained the heat shock element (HSE) in the promoter, indicating fast upregulation of this gene under elevated temperatures and possible role in heat stress response.

Posttranscriptional regulation of SOD genes expression includes RNA processing mechanisms such as microRNA-mediated silencing, alternative splicing and alternative polyadenylation. MicroRNAs (miRNAs) are a class of small non-coding RNA molecules (20-24 nucleotides), which participate in RNA silencing via base-pairing with complementary sequences within mRNA. One of the well-examined plant miRNAs is miR389. In *Arabidopsis*, miR389 interacts with mRNAs for two Cu/ZnSODs, CSD1 and CSD2. Formation of miRNA-mRNA duplexes leads to silencing of mRNA by degradation of the mRNA strand or repression of mRNA translation, which depends upon the degree of complementarity between miRNA and cognate mRNA (Kumar, 2014). Plant miRNAs usually have a near-perfect pairing with their mRNA targets, which results in cleavage and complete degradation of the target transcripts. The miR389 also indirectly controls the expression of CSD1 and CSD2 at posttranslational level through silencing of mRNA for CCS1, metallochaperone required for Cu delivery and full activation of Cu/ZnSODs (Beauclair et al., 2010). *A. thaliana* miR389 probably cooperates with AGO10 protein during miR398-directed translational inhibition of *CCS1* mRNAs (Beauclair et al., 2010). In general, miRNAs form a complex with Argonaute (AGO) protein which, as a part of RNA-induced silencing complex, guides miRNA to target its complementary mRNA sequence. The miR389 is also found in other plant species, including soybean, barrel clover and peanut (reviewed in Wang et al., 2016b), which indicates that miR389-mediated silencing might be rather widespread in plants. The miR389 expression is downregulated in response to cold stress, dehydration, salt stress and ABA (Sunkar and Zhu, 2004) which allows an increase in CSD1 and CSD2 transcript levels under abiotic stress conditions. The miRNA-mediated posttranslational regulation of SODs is predicted for upland cotton, *Gossypium hirsutum*, by comprehensive data search and analysis (Wang et al., 2017). Sequences of 20 miRNAs that could target 14 SOD genes at 33 prediction sites were detected in *G. hirsutum* genome. Target sites for *G. hirsutum* miR398 were found in coding regions of four Cu/ZnSOD genes: *GhCSD7*, *GhCSD8*, *GhCSD9* and *GhCSD10.* Interestingly, miRNA-mediated posttranscriptional regulation by ghr-miR3 was also postulated for four *G. hirsutum* MnSOD genes. *G. hirsutum* FeSOD genes *GhFSD3* and *GhFSD4* could be targeted by m0362 and novel_mir_4246, respectively, while *GhFSD1* and *GhFSD2* could be targeted by seven and more miRNAs with continuously distributed sites. Further investigation is needed to determine expression profiles of these miRNAs and their SOD targets, and give insight into miRNA biological functions in upland cotton (Wang et al., 2017). Apart from their role in abiotic stress-regulated expression of target SOD transcripts, miRNAs might have an important regulatory role in response to biotic stresses or developmental cues (Kumar, 2014).

Alternative splicing is a mechanism involved in the qualitative and/or quantitative control of gene expression by regulating mRNA stability and introducing functional diversity into the products of particular genes. In California poplar, *Populus trichocarpa*, analysis of the expressed sequence tag (EST) libraries has indicated the presence of two transcripts of high-isoelectric-point superoxide dismutase (PthipI-SODC1) that were designated as *hipI-SODC1b* and *hipI-SODC1s* (Srivastava et al., 2009). High-isoelectric-point superoxide dismutases (hipI-SODs) are isoforms of Cu/ZnSOD with substantially higher pI values than other known SODs (>9.5), which are detected in *Pinus* spp., *Zinnia*, and *Populus*. Alignment of transcripts' sequences with the *PthipI-SODC1* gene sequence revealed that *hipI-SODC1b* was 69 bp longer than *hipI-SODC1s* due to alternative splicing involving an alternative donor splice site in the sixth intron. Transcript analysis showed differential expression of splice variant *hipI-SODC1b* which was expressed in cambial and xylem, but not phloem. The *hipI-SODC1b* transcript encodes an inactive protein, while *hipI-SODC1s* transcript encodes an enzymatically active protein. It was suggested that the presence of additional amino acids at the C terminus might interfere with proper folding and dimerization of hipI-SODC1b, leading to inactivation of protein (Srivastava et al., 2009). Nonetheless, alternative splicing can also lead to novel, active SOD isoforms. In rice, *Oryza sativa* L. cv. Cpslo17, a novel alternatively spliced variant of the FeSOD gene transcript, *OsFeSODb*, was obtained from panicle cDNA library (Feng et al., 2006). The alternative splicing event happened by a fusion of exon 4, exon 5 and intron 4 of the *OsFeSOD* gene and led to a frameshift mutation, which caused a complete change in the amino acid sequence downstream of the Valine-169 residue. Two alternative forms of *OsFeSOD* transcripts, *OsFeSODa* and *OsFeSODb*, were expressed in both the vegetative and reproductive tissues of rice cultivar Cpslo17 as detected by RT-PCR analysis. Moreover, accumulation of both isoforms was up-regulated by light. SOD activity assay showed that C-terminal truncation in OsFeSODb did not result in a loss of enzyme activity. Authors also reported that the alternative splicing of *OsFeSOD* pre-mRNA was sensitive to low temperature. In Peruvian cotton, *Gossypium raimondii*, analysis of Cu/ZnSOD (GrCSD) and FeSOD (GrFSD) transcripts have revealed the existence of more than one transcript variant per gene. In particular, two transcripts were found for *GrCSD3*, and three transcripts for each of *GrCSD1*, *GrFSD1* and *GrFSD2*. An analysis of alternative splicing suggested that *GrCSD1*, *GrFSD1* and *GrFSD2* pre-mRNAs probably underwent three basic types of alternative RNA splicing events: intron retention, alternative 3'- or alternative 5' splice site (Wang et al., 2017).

Alternative polyadenylation has relatively recently been recognized as an important factor in gene regulation (Lutz, 2008). Pre-messenger RNAs (pre-mRNAs) may possess more than one polyadenylation signal and can be polyadenylated in several different ways, enabling a single gene to encode multiple mRNA transcripts. Although coding sequence remains unchanged, the difference in the 3'UTR length may influence mRNAs stability and translational efficiency by altering the availability of 3'UTR regulatory sequences, such as AU-rich elements (AREs) and miRNA-binding sites. Search for MnSOD cDNAs in *Zea mays* EST databases revealed numerous ESTs that can be divided into two large groups, corresponding to MnSOD3-1 and MnSOD3-4 cDNA (Katyshev et al., 2006a). Both EST groups show a substantial variability in the length of 3′-untranslated regions (3′-UTR) which could be explained by alternative polyadenylation of corresponding pre-mRNAs. In Dahurian larch, *Larix gmelinii* Rupr., one copy of MnSOD gene occurs in the genome as determined by sequencing of genomic DNA fragments and analysis of the 5'-terminal regions of the cDNAs (Katyshev et al., 2006b). The existence of several *L. gmelinii* MnSOD cDNAs, differing in the size of the 3'-terminal sequence, can be explained in terms of alternative polyadenylation (Katyshev et al., 2006a). Indeed, analysis of MnSOD cDNAs showed that the difference in their sizes was determined by a variable position of the poly(A) start site due to the existence of three canonic signals (AATAAA). In banana, 25 transcripts were transcribed from 12 different SOD genes (Feng et al., 2015). Transcripts encoding the same SOD shared an identical ORF, but differed in their UTRs. The authors revealed that besides alternative polyadenylation, alternative transcription start sites are responsible for observed abundance in SOD gene transcripts.

All plant superoxide dismutases are encoded by nuclear genes and SOD mRNAs are translated on 80S ribosomes in the cytosol. Newly synthesized organellar SOD polypeptides, no matter of the type of conformation acquired in the cytosol, are imported into organelles where final posttranslational processing occurs. For instance, chloroplastic FeSODs and Cu/ZnSODs possess cleavable chloroplast transit sequences that allow the transport of these proteins to the stroma via the TOC/TIC protein import machinery in the envelope, followed by cleavage of the transit sequence by the stromal transit sequence peptidase. Chloroplast import machinery has a strong unfolding capacity and proteins precursors are imported in an unfolded state (Jarvis, 2008). Therefore, final steps of posttranslational processing, which involve folding by chloroplast chaperones, cofactor assembly and subunit association, take place in the stroma. Maturation of Cu/ZnSODs also requires the aid of a metallo-chaperone CCS (Culotta et al., 2006). CCS delivers the Cu cofactor and assists in the formation of an intramolecular disulfide bridge required for the full activation of Cu/ZnSOD (Culotta et al., 2006). Metallochaperones for FeSOD and MnSOD assembly are not discovered so far, and it is not known how these enzymes acquire their metal cofactors.

Temporal and Spatial Specificity of SOD Expression

Comparison of the expression patterns of *Arabidopsis thaliana* superoxide dismutase over the entire life cycle, conducted *in silico*, shows that the major SOD isoforms are cytosolic CSD1, chloroplast CSD2 and cytosolic/chloroplastic FSD1 (Pilon et al., 2011). FSD1 is highly expressed in seedlings and younger *Arabidopsis* plants, while its expression significantly drops with the formation of rosettes and onset of flowering (Pilon et al., 2011). Conversely, the expression of cytosolic *CSD1* and chloroplast *CSD2* genes is relatively low at initial developmental stages and increases upon the onset of flowering (Pilon et al., 2011). Although chloroplastic *FSD2* and *FSD3* genes have a low basal expression level, its expression is slightly elevated at initial developmental stages resulting in more or less uniform expression level of chloroplast SODs, and probably activity, at all stages. FSD2 and FSD3 play essential roles in early chloroplast development in *Arabidopsis* (Myouga et al., 2008), and this could explain relatively high expression in the early developmental stages. The mitochondrial MnSOD gene, *MSD1*, is stably constitutively expressed at all developmental stages. Peroxisomal *CSD3* is expressed at much lower levels and only increases during flowering, consistent with the observation that the peroxisomal CSD3 is the principal SOD activity in flowers.

In rice, *Oryza sativa* L., the expression levels of cytosolic Cu/ZnSODs were high through the whole vegetative stage and dropped at the early flowering stage as detected by qRT-PCR analysis (Nath et al., 2014). Expression of rice chloroplastic *FeSOD2* and *FeSOD3* was very low in the early developmental stages and started to increase with age, which indicates differences in the expression pattern between rice and *Arabidopsis*. The expression of rice mitochondrial MnSOD was stable during the vegetative stage and dropped at flowering stage.

Modulation of SOD Expression by Metal Availability and Nutrition

Metal bioavailability is one of the factors that influence plant SOD expression at transcriptional, posttranscriptional and posttranslational level. In *Arabidopsis*, copper is required for expression of Cu/ZnSOD isoforms CSD1 and CSD2, as well as the Cu chaperone (CCS). The mechanism for the regulation by Cu involves Cu-responsive transcription factor SPL7 (Yamasaki et al., 2009). On low Cu, SPL7 induces biosynthesis of Cu microRNAs that down-regulate Cu enzymes. One of the Cu microRNAs is *miR398*, which interacts with mRNAs of CSD1, CSD2 and CCS and down-regulate expression of these Cu enzymes. Moreover, regulation of *Arabidopsis* FeSOD and Cu/ZnSOD gene expression is reciprocal in response to Cu bioavailability. Transcription factor SPL7 induces *FSD1* expression under low Cu. Higher Cu concentrations, however, promote expression of Cu/ZnSOD isoforms and reduce expression of FeSOD. In moss *Barbula unguiculata*, expression of the FeSOD gene was repressed under abundant Cu by binding of transcription factor PpSBP2 (SQUAMOSA promoter-binding protein) and its related proteins to GTACT motif (Nagae et al., 2008). The GTACT motif is a negative *cis*-acting element present in FeSOD gene, as well as in the promoter region of *miR398* gene, which seems to be essential in Cu sensing. Also, the moss FeSOD gene was negatively regulated by Cu in transgenic *Nicotiana tabacum* plants, and the *A. thaliana* FeSOD gene promoter containing the GTACT motif was repressed by Cu. These findings suggested that molecular mechanisms of GTACT motif-dependent *FeSOD* transcriptional suppression by Cu could be conserved in land plants (Nagae et al., 2008).

The addition of sucrose in the growth medium can affect Cu/ZnSOD expression in plants grown *in vitro*. Sucrose promoted miR398 expression and, consequently, caused down-regulation of *CSD1* and *CSD2* in *Arabidopsis* (Dugas and Bartel, 2008). Interestingly, FeSOD protein levels were slightly affected by sucrose, implying lack of regulation by SPL7 and indirectly suggesting that sucrose does not induce miR398 expression via SPL7.

Modulation of SOD Expression by Abiotic-Stress

Abiotic stresses modulate SOD expression at transcriptional and posttranscriptional level. Characterization of SOD gene families at the genome-wide level revealed the existence of numerous biotic/abiotic stress-responsive *cis*-elements responsible for regulation of gene transcription in SOD gene promoters (Feng et al., 2015; Feng et al., 2016; Wang et al., 2017; Zhou et al., 2017). The following types were detected: heat stress-responsive element (HSE), defense and stress-responsive element (TC-rich repeats), anaerobic induction element (ARE), drought-responsive element (MBS) and elicitor-responsive elements (EIRE and Box-W1). The frequent occurrence of drought-responsive element MBS, binding sites for MYB transcription factors, in SOD genes suggests an important role for the members of the MYB family in the regulation of SOD gene expression by environmental cues.

The MYB transcription factor family plays crucial roles in regulatory networks controlling development, metabolism and responses to biotic and abiotic stresses in plants (Dubos et al., 2010). The most common is the MYBR2R3 subfamily, encompassing proteins that possess two adjacent sequence repeats (R2 and R3) in MYB domain, with over 100 members in *Arabidopsis* and rice (Dubos et al., 2010). The R2R3 repeats of MYB transcriptional factors interact with the cognate site in the promoter region of genes and enable MYP protein binding which promotes gene transcription. In upland cotton, analysis of promoter regions of SOD genes revealed drought-responsive MBS element for the MYB-type transcription factor in all of 18 *GhSOD*s (Wang et al., 2017). The drought-responsive MBS elements were also found in 6 cucumber genes encoding Cu/ZnSOD, FeSOD and MnSOD, with the exception of *CsCSD1*, *CsCSD3* and *CsFSD2* (Zhou et al., 2017). In banana, *Musa acuminata* cv. Tianbaojiao, Cu/ZnSOD subfamily comprises six genes, *MaCSD1A*, *MaCSD1B*, *MaCSD1C*, *MaCSD1D*, *MaCSD2A* and *MaCSD2B* (Feng et al., 2015). The qPCR analysis of differential gene expression under cold, heat, drought or salinity treatment revealed that only *MaCSD1D* showed obvious expression changes under all four abiotic stresses, while other *MaCSD* genes were responsive to at least one abiotic stress treatment performed in the study. Compared with other *MaCSD* promoters, the *MaCSD1D* promoter harbored more kinds and numbers of *cis*-elements involved in abiotic stresses, including two MBS motifs (MYB binding sites) involved in drought-inducibility. The significance of particular MYB factors, as well as WRKY and ERF, in the modulation of SOD expression in genetically transformed plants is described in the later section of this chapter.

MicroRNA-mediated silencing is a posttranscriptional mechanism involved in modulation of SOD expression by abiotic stressors. So far, roles of two miRNAs, miR398 and miR408, in abiotic stress response were resolved in *Arabidopsis*. Downregulation of miR398 expression by oxidative stress increases posttranscriptional *CSD1* and *CSD2* mRNA accumulation and oxidative stress tolerance (Sunkar et al., 2006). Furthermore, transgenic *A. thaliana* plants overexpressing a miR398-resistant form of *CSD2* accumulated more *CSD2* mRNA than plants overexpressing a regular *CSD2*, and showed tolerant to heavy metals, high light, and other abiotic stressors (Sunkar et al., 2006). As described in previous section of this chapter, expression of CSD1 and CSD2, two Cu/ZnSODs important for abiotic stress response in *Arabidopsis*, is fine-tuned by miR398 and Cu bioavailability. Induction of the other miRNA, miR408, during abiotic stress reduces levels of the non-essential cuproproteins plantacyanin, cupredoxin, uclacyanin and LAC3 proteins and, thus, increases the availability of copper to those cuproproteins that are required to cope with the stresses, such as CSD1 and CSD2 (Ma et al., 2015). Indeed, transgenic *A. thaliana* plants with elevated miR408 expression were showing increased expression of CSD1 and CSD2, as well as Cu chaperone CCS, and improved tolerance to cold, salinity and methyl viologen-induced oxidative stress (Ma et al., 2015). It is reasonable to expect discovery of new miRNAs involved in abiotic stress-modulation of SOD expression in the future, considering number of miRNAs which target sites were predicted in *SOD* sequences by genome-wide approach (Wang et al., 2017).

PHYSIOLOGICAL ROLES OF SODS IN PLANTS UNDER ABIOTIC STRESS

Low-Temperature Stresses

Low-temperature (LT) stresses comprise cold (chilling temperatures, <20°C) stress and freezing (<0°C) stress which affect plant growth and development. Chilling-sensitive plants exposed to LT usually show water stress symptoms, such as reduction in the root water uptake and a decrease in root-to-leaf hydraulic conductance and turgor pressure, whereas tolerant plants have the capability to maintain conductance (Irigoyen et al., 1996; Aroca et al., 2001). Chilling leads to the overproduction of ROS in chloroplasts by disrupting ratio of absorbed and used light energy and by suppressing Calvin-Benson cycle (Logan et al., 2006, Zhou et al., 2006). Under such stress condition which limits CO_2 fixation, photosynthetic electron flux to O_2 will increase, causing overproduction of $O_2^{\bullet -}$, H_2O_2 and $OH^{\bullet}$ (Asada, 2006). Besides, mitochondria become an important site of ROS production under low temperature stresses as a result of over-reduction of the respiratory electron transport chain by low temperatures (Møller, 2001).

Responses to low temperature-induced oxidative stress include alterations in the activities of the ROS scavenging enzymes in plants. Liu et al. (2013) measured the activity of SODs in cold tolerant plant *Avena nuda* L. under two low-temperature treatments, 1°C and -10°C. At 1°C, activity of SODs increased rapidly in the first three days and remained at elevated level in later days. This significant increase in SODs activity indicates that *A. nuda* is resisting and adapting to chilling stress, and that SODs are contributing to reduced damage caused by cold (Liu et al.,2013). At -10°C, the SOD activity increased rapidly in the first three days and decreased later, probably due to observed extensive damage to seedlings that affected the SODs synthesis. These findings indicate that strong and rapid increase in SOD activity was not sufficient for the protection of *A. nuda* against freezing stress. In cucumber leaves, chilling temperature caused up-regulation of antioxidant enzymes, with a more significant SOD activity increase in chloroplast than in mitochondria (Hu et al., 2008). Also, chloroplast FeSOD showed a more significant increase in activity compared to chloroplast Cu/ZnSOD after chilling treatment at 8°C. Accumulation of chloroplast membrane-bound FeSOD, located close to the site of ROS production, might reduce the diffusion of ROS to photosystem II and cell membranes and alleviate damage to plant cell under cold stress (Hu et al., 2008). Furthermore, some results obtained from genetically modified plants indicate that MnSOD has a significant role in protecting plants against foliar damage and loss in biomass during LT treatments (Van Breusegem et al., 1999a; Samis et al., 2002). Vyas and Kumar (2005) reported that the induction of MnSOD in leaves during cold winter months improves winter survival of tea, *Camelia sinensis* L. O. Kuntze, plants. In addition, Xu et al. (2010) found difference in response to chilling stress between seedlings' roots and shoots in chilling tolerant tobacco variety Honghuadajinyuan. The activity of SODs in shoots was similar to the control under chilling stress, while the activity of SODs increased in roots. Interestingly, the relative growth rate in the chilling period compared to recovery period was significantly higher in roots than that in shoots for this tobacco cultivars, suggesting that root growth was the less affected by chilling stress. In leaves of *Nicotiana tabacum* cv. Petit Havana SR1, the activity of all antioxidant enzymes except SOD significantly decreased at 5°C compared to enzyme activity at 25°C (Gechev et al., 2003), which reflects SOD functional stability under LT.

High-Temperature Stress

High-temperature stress or heat stress has an adverse impact on plant growth and development. Extremely high temperatures and prolonged heat waves, which severely reduce crop quality and productivity, will become more frequent in the future due to global warming. The chloroplasts and mitochondria are the major sources of ROS in plant cells under high temperatures. Oxygen is photo produced by splitting water at photosystem II can form superoxide anion radical ($O_2^{\bullet-}$) through univalent reduction by electrons emerging from ferredoxin in photosystem I (Asada et al., 1974). To cope with oxidative damage, chloroplasts employ two antioxidant enzymes, Cu/ZnSOD and APX that function synergistically to enhance response to oxidative stress (Kwon et al., 2002). In spinach chloroplasts, over one-half of Cu/ZnSOD is bound to the stromal-faces of thylakoid membranes at or near PSI complex (Ogawa et al., 1995).

Almeselmani et al. (2006) conducted the field experiment on five wheat genotypes, PBW343, PBW 175, HDR-77, HD 2815 and HD 2865, to study the effect of high temperatures on the antioxidant enzymes activity. Two of these genotypes were heat tolerant (HDR77 and HD2815), the other two were moderately susceptible (PBW175 and HD2865), and PBW343 was susceptible to high temperatures. Heat tolerant genotypes HDR77 and HD2815 have relatively higher activity of SODs and the other investigated antioxidant enzymes compared to PBW 343, PBW 175 and HD 2865 at all stages of plant growth after late and very late planting (higher cumulative averages of daily temperature). Also, HDR77 and HD2815 showed minimum reduction in chlorophyll content and lower membrane injury index compared to other genotypes which indicate that oxidative stress induced by high temperature was ameliorated by ROS-scavenging enzymes (Almeselmani et al., 2006). Almeselmani et al. (2009) reported that heat-tolerant wheat cultivar C306 showed higher activity level of SODs and other antioxidant enzymes relative to heat-sensitive PBW343 at different growth stages during heat stress treatment. Recently, Zou et al. (2017) investigated effects of high temperature on the growth and development of two *Brassica campestris* cultivars, WS-6 and WS-1, with opposite response to heat-stress treatments. Results from the study revealed that cv. WS-1 exhibits smaller decreases in growth and photosynthetic parameters under heat stress compared to cv. WS-6. Also, transmission electron microscopy revealed smaller damage to the photosynthetic apparatus in thylakoid membranes of WS-1 cells compared to cv.WS-1 under heat stress. These negative effects of high temperature on WS-6 were probably the result of a greater accumulation of reactive oxygen species (ROS) and malondialdehyde (MDA). Heat-tolerant cultivar WS-1 showed higher activity level of SODs and other ROS-scavenging enzymes compared to WS-6.

Drought Stress

Drought stress may be considered to be one of the most important abiotic stresses with the greatest impact on plant growth and development, productivity, and survival (Jacobsen et al., 2012). This abiotic stress also determines the natural distribution of plant species. Water stress can simultaneously occur during the course of other environmental stresses, such as salt stress and cold stress. Drought stress adversely impacts many aspects of a plant's physiology in which photosynthesis is one of the primary physiological targets (Flexas et al., 2004), leading to higher production of ROS (Miller et al., 2010). For instance, rice seedling subjected to polyethylene glycol (PEG)-mediated drought (osmotic) stress for 24 h at -0.5 and -2.0 MPa showed between 114 and 167% increase in the concentration of $O_2^{\bullet-}$ compared to control (Sharma end Dubey, 2005). These findings suggested that drought stress induces oxidative stress in rice. The number of studies reported effects of drought stress on the expression and activities of the key enzymes involved in scavenging and detoxification of ROS (Malan et al., 1990; Sofo et al., 2005; Liu et al., 2014). Perl-Treves and Galun (1991) followed expression of the two Cu/ZnSOD genes encoding chloroplast and cytosolic Cu/ZnSOD in tomato plants and reported that the level of the cytosolic transcript and the respective activity of isozyme significantly increased during prolonged drought stress condition, while the abundance of chloroplast transcript remained unaffected.

The significance of SODs in drought response was also investigated in several drought-tolerant species. The drought-tolerant star moss, *Tortula ruralis*, showed higher levels of Cu/ZnSOD enzyme activity and reduced level of lipid peroxidation compared to the drought-sensitive fern-leaved moss, *Cratoneuron filicinum* (Dhindsa and Matowe, 1981). *Caragana* genus comprises species of desert shrubs with prominent drought-tolerance. Zhang et al. (2014a) isolated full-length sequences of Cu/ZnSOD genes from three *Caragana* species, *C. korshinskii*, *C. microphylla*, and *C. intermedia*, using degenerate polymerase chain reaction (PCR) and rapid amplification of cDNA ends (RACE) techniques. The cloned SOD cDNAs of *C. korshinskii* and *C. intermedia* shared 100% sequence identity, while a 24-bp specific sequence was found in the 3-UTR region of *C. microphylla* Cu/Zn-SOD cDNA. Expression of three identified Cartagana *Cu/ZnSOD* genes was examined *in vitro* by PEG-mediated drought stress. All three *Cu/ZnSOD* were induced by PEG-simulated drought stress and ABA, and showed similar expression patterns, with no significant differences in transcriptional levels among them. These findings suggested the importance of Cu/ZnSOD isoform in the drought-stress response of *Caragana* spp.

Salt Stress

High concentrations of salts in soil or growing medium cause both osmotic stress and ionic toxicity, leading to overproduction of ROS and oxidative stress in plants. Pan et al. (2006) studied the effect of salt stress on the antioxidant enzyme system in liquorice seedlings (*Glycyrrhiza uralensis Fisch*). The results showed that salt stress upregulated activities of SODs. Under the treatment of 2% NaCl, the activity of MnSOD significantly increased, FeSOD activity was not affected, while activity CuZnSOD declined in seedlings. Also, the authors found induction of a novel MnSOD isoform in liquorice subjected to 2% NaCl for 48, 72 and 96 h. These findings suggested the existence of MnSOD-gene family, with members that show differential expression under salt stress. Induction of mitochondrial MnSOD (mtMnSOD) in response to salt stress was previously detected in pea, *Pisum sativum* L. cv. Puget, plants (Gomez et al., 1999). Results revealed that activity of mtMnSOD was significantly induced only under severe NaCl concentration (110-130 mM). In chloroplasts, main SOD activity could be attributed to Cu/ZnSODII (1.7-fold increase under severe salt stress), although FeSOD exhibited a stronger enhancement (3.5 fold) in its activity compared to Cu/ZnSOD in pea exposed to high salt concentration. Rice is a crop of great economic importance in the world and also one of the most sensitive plants to the high-salt environment. Accumulation of salts in irrigated soil is one of the primary causes that limit production of this crop species in Southeast Asia. To determine the significance of SODs in response to salt stress in the rice, *Oryza sativa* L. cv. Dongjin, plants, Lee et al. (2001) investigated activities of these enzymes. The authors detected an increase in SODs activity, as well as the higher H_2O_2 content in rice leaves under salinity treatment compared to the control. Results from the native-PAGE analysis showed that rice plants possess two isoforms of MnSOD and five isoforms of Cu/ZnSOD in the leaves, while FeSOD isoforms were not detected. Activities of three Cu/ZnSOD isoforms, labeled 3, 4 and 5, were barely detected in leaves of control plants by native gels, while salt stress caused a significant increase in the activity of Cu/ZnSOD isoforms 1 and 2. Furthermore, the activity level of only one MnSODs, isoform 2, increased in leaves under salt stress. Also, salt stress enhanced the content of H_2O_2 as well as the activities of ascorbate peroxidase (APX), and guaiacol peroxidase, whereas it induced the decrease of catalase activity and had little effect on the activity levels of glutathione reductase (GR). The authors suggested that overproduction of H_2O_2, by enhanced activity of SODs, functions as the signal of salt stress, which induces the induction of specific APX isoforms but not specific GR isoforms.

Hypoxia

Low molecular oxygen concentration or hypoxia is an environmental challenge that plants encounter throughout the entire lifecycle. Hypoxia can occur not only during natural events (e.g., winter ice, spring floods, the excess of rainfall), but also during a developmental process (seed germination) or a dramatic reduction of photosynthetic efficiency (Kennedy et al., 1992; Else et al., 2009). Oxygen is of vital importance for plant energy metabolism. Since plants do not have a mechanism to deliver oxygen to different tissues, a plant cell has to continuously struggle with changes in the oxygen concentration. To understand the mechanism of hypoxia stress response, plant species that range from submergence sensitive to submergence tolerant have been investigated. Wetland species such as rice (Van der Straeten et al., 2001), *Ranunculus pygmaeus* (Horton, 1992), *Rumex palustris* (Blom et al., 1994) and *Polygonum lapathifolium* (Carter and Grace, 1990) can survive complete submergence and they are excellent model plants for studying reactions upon flooding. On the other hand, *Arabidopsis*, wheat, barley and maize are sensitive to complete submergence, and they are suitable for investigating effects of flooding on root systems (Fukao and Bailey-Serres, 2004). Although low light condition during submergence decreases ROS generation, waterlogging induces high ROS production (Kumutha et al., 2009; Roach and Krieger-Liszkay, 2014). Thus, antioxidative enzymes are needed for ROS scavenging. In rice, abundance of SODs in cell organelles differs between seedlings germinated under water and standard conditions (Ushimaru et al., 1999). The level of different SOD isoforms in rice cells reflects the development of corresponding organelles. Under submergence, plastids are undeveloped and developed plastids degrade rapidly, so the level of plastidic Cu/ZnSOD is low and notably drops. Conversely, mitochondria are functional organelles and mitochondrial MnSODs are abundant (Ushimaru et al., 1999). The maintenance of MnSOD in rice seedlings germinated under water might have an important role in the rapid scavenging of ROS generated from mitochondria after hypoxia condition, while degenerated plastids probably do not need to produce sufficient amount of Cu/ZnSOD for protection from ROS during the post-hypoxic phase. Investigation on barley, however, showed that hypoxia does not affect MnSOD activity or induces synthesis of different MnSOD isoenzymes in roots (Szal et al., 2004), which can be explained by the difference in stress response between one wetland species and one waterlogging-sensitive plant species. In *Brassica napus*, hypoxia substantially reduces the activity of SODs in seedling leaves (Zhou et al., 1997). Decreased activity of SODs was also observed in maize leaves after seven days of flooding, especially in younger leaves (Yan et al., 1996). Different responses in SODs production and activity among plants rely not only on ROS levels, but whether changes in ROS levels cause damage or contribute to signaling in hypoxic plants.

Genetic Engineering of Plant SODs

Some of the earliest evidence for SOD involvement in abiotic stress responses came from enzyme activity and transcription analyses. Genetic engineering approach provided tools for more direct functional analysis of SODs, as well as for the fast and reliable creation of stress-tolerant plants. In recent years, stress-tolerant transgenic plants have been obtained in many plant species using transgenic technology with SOD genes (Table 1.).

Gupta et al. (1993) developed transgenic tobacco plants overexpressing *Pisum sativum* Cu/ZnSOD II which showed a 3.5-fold higher SOD activity than wild-type plants. In another study, *Pisum sativum* MnSOD-transformed rice (*Oryza sativa*) exhibited 1.4-fold higher total SOD activity compared with non-transgenic, which caused an increase in resistance to drought conditions (Wang et al., 2005). Also, Wang et al. (2010a) introduced a gene for MnSOD from *Tamarix androssowii*, *TaMnSOD*, driven by the CaMV35S promoter, into poplar (*Populus davidiana* x *Populus bolleana*). In transgenic poplar, the activity of SOD was significantly higher compared to non-transformed plants both before and after NaCl stress. Shiraya et al. (2015) generated transgenic rice plants with either constitutive high expression or suppression of novel Golgi/plastid-type MnSOD1, MSD1. Quantitative shotgun proteomic analysis indicated that the overexpression of *MSD1* up-regulated scavengers of reactive oxygen species under heat stress, and it was proposed that the Golgi/plastid MSD1 plays an important role in adaptation to high temperature conditions (Shiraya et al., 2015). These results demonstrate that the genetic manipulation with single SOD gene might improve tolerance to abiotic stress in the transgenic plants.

An important strategy in overexpressing SOD genes in plants for abiotic stress tolerance is to determine the isoform of SODs and the cell organelle for the overexpression (Bowler et al., 1992). It has been shown that overexpressing of SOD genes in the chloroplasts rather than in the cytosol is a more effective approach in protecting plants against induced biocide (methyl viologen) mediated oxidative damage (Perl et al., 1993). Prashanth et al. (2008) have increased tolerance to drought and salt stresses in an *indica* rice variety Pusa Basmati-1 by overexpressing of *SOD1* encoding a cytosolic Cu/ZnSOD from the mangrove, *Avicennia marina* Forsk. Badawi et al. (2004) transformed tobacco plants with chimeric gene consisting of the coding sequence for cytosolic *Oryza sativa* Cu/ZnSOD fused to the chloroplast transit sequence from *Arabidopsis* glutathione reductase. In this study, transgenic tobacco plants showed enhanced tolerance to salt stress and water deficit over the wild-type, suggesting that overexpressed of rice Cu/ZnSOD increase chloroplast antioxidant defense. Besides, the transformed tobacco exhibited higher photosynthetic performance during photo-oxidative stresses such as high salt, drought, and PEG treatment, compared to non-transgenic plants. Another example of the importance of cytosolic Cu/ZnSOD for increasing tolerance to abiotic stress was provided by Pal et al. (2013). The authors transformed potato (*Solanum tuberosum* ssp. *tuberosum* L. cv. Kufri Sutlej) with a cytosolic *Cu/ZnSOD*, cloned from *Potentilla atrosanguinea* (*PaSOD*), which resulted in a high level of Cu/ZnSOD in the cytosol and enhanced tolerance to drought stress. In transgenic *Arabidopsis*, MnSOD overexpression and accumulation in mitochondria was found to enhance salt tolerance (Wang et al., 2004). Furthermore, transgenic rice plants overexpressing *OsMSD* in mitochondria showed the higher efficiency of $O_2^{\bullet -}$ elimination under various stress conditions (Li et al., 2013).

To determine importance of *FeSOD* overexpression in chloroplast for an increase in ROS scavenging capacity and improvement in the winter survival of alfalfa plants, two *Medicago sativa* L. genotypes were transformed with the vector pEXSOD10 (McKersie et al., 2000). The vector contained a cDNA for *Arabidopsis* FeSOD with a chloroplast transit peptide and cauliflower mosaic virus 35S promoter (CaMV35S). Transgenic alfalfa plants have shown improvement in winter survival, reduced secondary injury symptoms and enhanced recovery, which supported McKersie et al. (2000) hypothesis that perennial and winter annual plants experience oxidative stress during winter, and that antioxidant defense systems contribute to winter hardiness.

In contrast to the findings mentioned above, some transgenic plants with *SOD* overexpression did not exhibit improvements to oxidative or environmental stresses (Pitcher et al., 1991; Van Camp et al., 1996; Payton et al., 1997). Earlier, Tepperman and Dunsmuir (1990) showed that transgenic tobacco (*Nicotiana tabacum*) plants transformed with *Cu/ZnSOD* from petunia appear to be no more oxidative stress-tolerant than wild-type plants. Another example was provided by Van Breusegem et al. (1999b). In their study, overproduction of *Arabidopsis thaliana* FeSOD in transgenic maize, *Zea mays* L., did not enhance chilling tolerance, but transgenic plants possessed better growth characteristics relative to non-transformed. These contradictory findings may be due to the complexity of plant antioxidant defenses mechanisms, which includes cooperation between a number of enzymes, and indicate that an increased tolerance to oxidative stress could be achieved by introduction of several antioxidant defense genes, i.e., gene pyramiding into a single genotype (Halpin, 2005). Considering that realization of this concept in practice is sometimes complicated, several approaches, such as retransformation, co-transformation, and the use of ‘linked effect genes’, have been developed to avoid pitfalls (Halpin et al., 2001; Halpin, 2005; Cingel et al., 2014). Some authors were testing these approaches to introduce more than one SOD-encoding transgene in plants in order to overproduce SODs and enhance plant tolerance to different stresses. Results from the study by Samis et al. (2002) on alfalfa indicate little or no improvement when two copies of SOD transgenes, Mit-MnSOD and the Chl-MnSOD transgene, were simultaneously expressed in the F1 progeny. Siblings expressing only one or the other transgene had higher shoot and storage organ biomass, and improved persistence. The authors postulated that usage of the promoter with lower expression than 35S might be a better solution for pyramiding of closely related transgenes.

Co-expression of two different antioxidant enzymes resulted in significant enhancement in stress tolerance (Kwon et al., 2002; Badawi et al., 2004; Faize et al., 2011; Shafi et al., 2015; Shafi et al., 2017). For instance, simultaneous overexpression of cassava Cu/ZnSOD and pea ascorbate peroxidase (APX) in transgenic tall fescue, *Festuca arundinaceae* Schreb. cv. Kentucky-31, increased tolerance to several ROS-generating abiotic stresses (Lee et al., 2007). Luo et al., (2013) revealed that simultaneous overexpression of two genes from *Gossypium hirsutum*, encoding SOD (*GhSOD1*) and catalase (*GhCAT1*), shows the synergistic effect of enhancing tolerance to methyl viologen and salt stress in transgenic cotton plants. Recent work by Shafi et al. (2017) demonstrated that transgenic potato overexpressing genes for both Cu/ZnSOD (*PaSOD*) from *Potentilla atrosanguinea* and APX (*RaAPX*) from *Rheum australe* exhibits a higher tolerance to severe saline conditions than non-transformed plants. All these results indicate that the combination of multiple antioxidative enzymes transgenes could have a synergistic effect on stress tolerance in plants and could be more convenient to enhance stress tolerance than the use of a single transgene.

Table 1. List of representative studies on the abiotic stress-tolerant transgenic plants produced by overexpressing SODs

Protein	Transgene	Origin	Transformed plant	Abiotic stress tolerance	Reference
Cu/ZnSOD	*AhCu/ZnSOD*	*Arachis hypogaea*	*Nicotiana tabacum* cv. Xanthium	Salinity and drought	Negi et al., (2015)
	OsCu/Zn-SOD	*Oryza sativa* var. Longjing11	*Oryza sativa* var. Longjing11	Salinity	Guan et al., (2017)
	KcCSD	*Kandelia candel*	*Nicotiana tabacum*	Salinity	Jing et al., (2015)
	SiCSD	*Saussurea involucrate* Kar. et Kir.	*Nicotiana tabacum* L. cv. NC89	Drought and cold	Zhang et al., (2017)
	Cu/ZnSOD	*Oryza sativa*	*Nicotiana tabacum*	Salinity, water deficit and PEG	Badawi et al., (2004)
	sod1	*Avicennia marina*	*Oryza sativa* var. Pusa Basmati-1	Salinity	Prashanth et al., (2008)
	JcCu/ZnSOD	*Jatropha curcas* L.	*Arabidopsis thaliana*	Salinity	Liu et al., (2015)
FeSOD	*FeSOD*	*Arabidopsis thaliana*	*Zea mays* L.	Chilling	Van Breusegem et al., (1999)
	FeSOD	*Arabidopsis thaliana*	*Medicago sativa* L.	Freezing	McKersie et al., (2000)
MnSOD	*WMnSOD1*	*Triticum aestivum*	*Brassica napus* cv. Westar	Aluminium	Basu et al., (2001)
	pMnSOD	*Pisum sativum*	*Oriza sativa*	Methyl viologen and Drought	Wang et al., (2005)
	TaMnSOD	*Tamarix albiflonum*	*Gossypium hirsutum L.*	Drought	Zhang et al., (2014b)
	MnSOD	*Nicotiana plumbaginifolia*	*Medicago sativa*	Drought	McKersie et al., (1996)
	TaMnSOD	*Tamarix androssowii*	*Populus davidiana* × *P. bolleana*	Salinity	Wang et al., (2010a)

Table 1. (Continued)

Protein	Transgene	Origin	Transformed plant	Abiotic stress tolerance	Reference
Cu/ZnSOD & catalase	*Cu/ZnSOD + CAT*	*Zea mays* L.	*Brassica campestris* L. ssp *pekinensis* cv. Tropical Pride	Salinity	Tseng et al., (2007)
	GhSOD1 + GhCAT1	*Gossypium hirsutum*	*Gossypium hirsutum*	Methyl viologen and salinity	Luo et al., (2013)
	MeCu/ZnSOD + MeCAT	*Manihot esculenta* Crantz.	*Manihot esculenta* Crantz.	Methyl viologen and chilling	Xu et al., (2014)
Cu/ZnSOD & ascorbate peroxidase	*Cu/ZnSOD + APX*	SOD – *Manihot esculenta* APX – *Pisum sativum*	*Festuca arundinacea* Schreb. cv. Kentucky-31	Methyl viologen and heavy metals	Lee et al., (2007)
	PaSOD +RaAPX	SOD – *Potentilla atrosanguinea* APX – *Rheum australe*	*Arabidopsis thaliana*	Salinity	Shafi et al., (2015)
	cytsod + cytapx	*Spinacia oleracea*	*Prunus domestica* cv. Claudia Verde	Salinity	Diaz-Vivancos et al., (2013)
	mSOD1 + APX	SOD – *Manihot esculenta* APX – *Pisum sativum*	*Ipomoea batatas* Lam.	Methyl viologen and chilling	Lim et al., (2007)
	AhCu/ZnSOD + AhcAPX	*Arachis hypogaea*	*Nicotiana tabacum* cv. Xanthi	Salinity	Negi et al., (2016)
MYB	*TaODORANT1*	*Triticum aestivum* cv. Chinese Spring	Tobacco	Drought and salinity	Wei et al., (2017)
	GbMYB5	*Gossypium barbadense*	*Nicotiana tabacum*	Drought	Chen et al., (2015)
	SoMYB18	Sugarcane var. Co740	Tobacco	Drought, salinity, cold	Shingote et al., (2015)
	BplMYB46	*Betula platyphylla*	*Betula platyphylla*	Salinity	Guo et al., (2017)
WRKY	*GhWRKY6-like*	*Gossypium hirsutum* cv. Jin668	*Arabidopsis thaliana* ecotype Columbia	Salinity	Ullah et al., (2017)
	GhWRKY39	*Gossypium hirsutum*	*Nicotiana benthamiana*	Salinity	Shi et al., (2014)

Protein	Transgene	Origin	Transformed plant	Abiotic stress tolerance	Reference
	GhWRKY41	*Gossypium hirsutum*	*Nicotiana benthamiana*	Drought and salinity	Chu et al., 2015
	SpWRKY1	*Solanum pimpinellifolium* L3708	*Nicotiana tabacum*	Drought	Li et al., (2015)
	TaWRKY44	*Triticum aestivum* cv. Chinese Spring	Tobacco	Drought and salinity	Wang et al., (2015)
	BdWRKY36	*Brachypodium distachyon*	*Nicotiana tabacum* L. cv. Samsun	Drought	Sun et al., (2015)
ERF	*OjERF*	*Ophiopogon japonicus*	*Nicotiana tabacum*	Drought	Li et al., (2012)
	JERF3	*Solanum lycopersicum*	*Nicotiana tabacum*	Salinity, drought and freezing	Wu et al., (2008)
	CaPF1	*Capsicum annuum*	*Pinus virginiana* Mill.	Heavy metals and heat	Tang et al., (2005)
bZIP	*GhABF2*	*Gossypium hirsutum* L.	*Gossypium hirsutum* L.	Drought and salinity	Liang et al., (2016)
	ThbZIP1	*Tamarix hispida*	Tobacco	Salinity	Wang et al., (2010b)
	PtrABF	*Poncirus trifoliate* L. Raf.	Tobacco	Drought	Huang et al., (2010)
Zinc finger	*ZFP245*	*Oryza sativa* L. sub. Japanonic cv. Zhonghua11	*Oryza sativa* L. sub. Japanonic cv. Zhonghua11	Drought	Huang et al., (2009)
	ThZFP1	*Tamarix hispida*	*Tamarix hispida*	Salinity	Zang et al., (2015)
	ZFP179	*Oryza sativa L.* subsp. *japonica* cv. Jiucaiqing	*Oryza sativa L.* subsp. *japonica* cv. Zhonghua11	Salinity and drought	Sun et al., (2010)
	AtRZFP	*Arabidopsis thaliana*	*Arabidopsis thaliana*	Salinity and osmotic	Zang et al., (2016)
	GhTZF1	*Gossypium hirsutum* cv. YZ1	*Arabidopsis thaliana*	Drought	Zhou et al., (2014)

Overexpression of SODs and/or high SODs activity in some transgenic plants might also be achieved by transformation with genes encoding transcription factors. Particular transcription factors are known to regulate transcription of more than a few genes involved in abiotic stress response and, thus, by simultaneously changing the expression of these genes, could enhance plant tolerance to unfavorable environmental conditions. One group of the transcriptional factors with an important role in abiotic stress responses is MYB superfamily. In tobacco, a higher level of *SODs* expression was detected in transgenic plants that overexpressed cotton MYB transcription factor, *GbMYB5*, compared to wild-type under drought stress (Chen et al., 2015). Besides, the transcript levels of the antioxidant genes CAT and GST, polyamine biosynthesis genes, drought-responsive genes *NCED3*, *BG* and *RD26* and the late embryogenesis abundant protein-encoding gene *ERD10D* were also higher in transgenic tobacco than in the wild-type plants under drought stress. This study suggested that *GbMYB5*-overexpression could enhance tobacco tolerance to the oxidative stress caused by drought conditions through activating the expression of a number of stress-related genes, including genes for antioxidative enzymes SOD, CAT and GST, genes regulating biosynthesis of osmolytes and LEA proteins and dehydration-responsive genes in the ABA-dependent signaling pathway. In the study on rice, R2R3-type MYB gene *OsMYB2* was isolated and functionally characterized (Yang et al. 2012). The OsMYB2 was localized in the nucleus with transactivation activity. Expression of OsMYB2 was up-regulated by salt, cold, and dehydration stress. OsMYB2 role in tolerance to abiotic stress was further investigated by generating transgenic rice plants with overexpressing and RNA interference OsMYB2. The *OsMYB2*-overexpressing and wild-type plants were not differing in growth and development under normal growth conditions. The *OsMYB2*-overexpressing plants, however, were more tolerant to salt, cold, and dehydration stresses and more sensitive to abscisic acid than wild-type plants. The *OsMYB2*-overexpressing plants were also showing the enhanced activities of SOD, as well as POD and CAT, which is the probable reason for markedly lower ROS and malondialdehyde (lipid peroxidation marker) content detected under stress conditions. Recent investigation by Wei et al. (2017) supports these findings of enhanced expression of a number of stress-related genes by single, overexpressed MYB gene under drought stress. The authors cloned MYB transcriptional factor *TaODORANT1*, a R2R3-MYB gene from wheat, and introduced the recombinant gene into tobacco. Transgenic, *TaODORANT1*-overexpressing tobacco plants exhibited lower Na^+ accumulation in leaves under salt stress, as well as higher relative water content and lower water loss rate under drought compared to non-transformed. The transcription level of *SODs*, examined by qRT-PCR after drought stress, was higher in transgenic tobacco than wild-type. Besides, *TaODORANT1* overexpression up-regulated the expression of several other ROS- and stress-related genes in response to both drought and salt stresses, thus enhancing transgenic tobacco plants tolerance.

WRKYs are another group of relevant transcriptional factors involved in the stress-related expression of *SODs*. Chu et al. (2015) isolated *Gossypium hirsutum* WRKY gene *GhWRKY41* and used it for the transformation of *Nicotiana benthamiana*. *GhWRKY41*-overexpression in response to osmotic stress enhanced tolerance of transgenic tobacco. Transgenic plants exhibited higher SODs enzyme activity than wild-type. The expression of SODs was also upregulated in transgenic plants exposed to drought and salt stress (Chu et al., 2015). Sun et al. (2015) showed that WRKY gene isolated from *Brachypodium distachyon*, *BdWRKY36*, acts as the positive regulator by controlling ROS homeostasis and upregulating transcription of *SODs* in transgenic tobacco lines under drought condition. Transcriptional factor ERF (ethylene response factor) was also demonstrated to be an important player in the regulation of transcription and activity of SODs under abiotic stress. For instance, overexpression of *JERF3*, initially isolated from tomato plants and transformed into tobacco, enhance ROS scavenging and increase tolerance of transgenic plants to several abiotic stress conditions (Wu et al., 2008). The authors reported that overexpression of *JERF3* enhanced activity of SODs by 1.6 – 2.1-fold in transgenic tobacco compared to wild-type plants after dehydration, cold, salt, and ABA treatments. This finding indicates the importance of JERF in the possible regulation of SOD genes expression at the transcriptional level. Besides mentioned, bZIP (basic leucine zipper) (Wang et al., 2010b; Huang et al., 2010; Liang et al., 2016) and zinc finger (Sun et al., 2010; Zhou et al., 2014; Zang et al., 2015; Zang et al. 2016) transcription factors have also been confirmed to upregulate expression of *SODs* and positively regulates response to oxidative stress in transgenic plants.

CONCLUSION

Our knowledge of plant SODs structure, function and gene regulation is likely to improve substantially in the future. The whole-genome sequences of many plant species were made available to the public, facilitating molecular studies. New disciplines have been developed in the postgenomic era, such as transcriptomics, proteomics, comparative and functional genomics, which provide tools for better understanding of SOD enzymes. Current characterizations of SOD gene families at the genome-wide level could help in determining full complements of SOD-encoding genes in various plant species and give insight into gene structure and possible molecular mechanisms of SOD genes regulation.

The investigations of SODs using a transgenic approach have shown that overexpression of single SOD gen can enhance plant stress tolerance in some instances. However, different SOD isoforms of the same type frequently have distinct subcellular localization and can be involved in various stress responses. Targeting of overexpressed SOD isoforms to different cellular compartments might be considered as a more effective approach for enhancement of plants stress response. Pyramiding of SOD recombinant genes, comprising SOD isoforms-coding sequence under control of moderate promoters to avoid hyperexpression, into a single genotype might result in the synergistic action.

Besides being an important antioxidant enzyme, SOD represents just one component of the complex antioxidant defense system in the plant cell. Modulation of SOD expression can disturb balance between the level of SOD activity and activity of other antioxidant enzymes, as well as impair cooperation between enzymatic and non-enzymatic components of the antioxidant system. Pyramiding of several antioxidant enzyme-genes, such as SOD and APX or CAT, might be considered as a promising strategy for the enhancement of crops' response against various environmental stresses. Future effort should be directed towards simultaneous modulation of the activity of both enzymatic and non-enzymatic components of plant antioxidant defense system in plants. Identification and application of particular transcription factors could help in the synchronous activity modulation of these components, as well as induction of additional stress-related proteins to improve plant abiotic stress tolerance.

ACKNOWLEDGMENTS

The authors would like to acknowledge financial support from Ministry of Education, Science and Technological Development, Republic of Serbia, Project Grant no. TR31049.

REFERENCES

Abreu, I. A. & Cabelli, D. E. (2010). Superoxide dismutases - a review of the metal-associated mechanistic variations. *Biochimica et Biophysica Acta (BBA)-Proteins and Proteomics*, *1804*, 263-274.

Allen, M. D., Kropat, J., Tottey, S., Del Campo, J. A. & Merchant, S. S. (2007). Manganese deficiency in *Chlamydomonas* results in loss of photosystem II and MnSOD function, sensitivity to peroxides, and secondary phosphorus and iron deficiency. *Plant Physiology*, *143*, 263-277.

Almeselmani, M., Deshmukh, P. S., Sairam, R. K., Kushwaha, S. R. & Singh, T. P. (2006). Protective role of antioxidant enzymes under high temperature stress. *Plant Science*, *171*, 382-388.

Almeselmani, M., Deshmukh, P. & Sairam, R. (2009). High temperature stress tolerance in wheat genotypes: role of antioxidant defence enzymes. *Acta Agronomica Hungarica*, *57*, 1-14.

Alscher, R. G., Erturk, N. & Heath, L. S. (2002). Role of superoxide dismutases (SODs) in controlling oxidative stress in plants. *Journal of Experimental Botany*, *53*, 1331-1341.

Amano, A., Shizukuishi, S., Tamagawa, H., Iwakura, K., Tsunasawa, S. & Tsunemitsu, A. (1990). Characterization of superoxide dismutases purified from either anaerobically maintained or aerated *Bacteroides gingivalis*. *Journal of Bacteriology*, *172*, 1457-1463.

Aroca, R., Tognoni, F., Irigoyen, J. J., Sánchez-Díaz, M. & Pardossi, A. (2001). Different root low temperature response of two maize genotypes differing in chilling sensitivity. *Plant Physiology and Biochemistry*, *39*, 1067-1073.

Asada, K., Kiso, K. & Yoshikawa, K. (1974). Univalent reduction of molecular oxygen by spinach chloroplasts on illumination. *Journal of Biological Chemistry*, *249*, 2175-2181.

Asada, K. (2006). Production and scavenging of reactive oxygen species in chloroplasts and their functions. *Plant Physiology*, *141*, 391-396.

Badawi, G. H., Yamauchi, Y., Shimada, E., Sasaki, R., Kawano, N., Tanaka, K. & Tanaka, K. (2004). Enhanced tolerance to salt stress and water deficit by overexpressing superoxide dismutase in tobacco (*Nicotiana tabacum*) chloroplasts. *Plant Science*, *166*, 919-928.

Basu, U., Good, A. G. & Taylor, G. J. (2001). Transgenic *Brassica napus* plants overexpressing aluminium-induced mitochondrial manganese superoxide dismutase cDNA are resistant to aluminium. *Plant, Cell & Environment*, *24*, 1278-1269.

Beauclair, L., Yu, A. & Bouché, N. (2010). microRNA-directed cleavage and translational repression of the copper chaperone for superoxide dismutase mRNA in Arabidopsis. *The Plant Journal*, *62*, 454-462.

Blom, C. W. P. M., Voesenek, L. A. C. J., Banga, M., Engelaar, W. M. H. G., Rijnders, J. H. G. M., Van de Steeg, H. M. & Visser, E. J. W. (1994). Physiological ecology of riverside species: adaptive responses of plants to submergence. *Annals of Botany*, *74*, 253-263.

Bowler, C., Montagu, M. V. & Inze, D. (1992). Superoxide dismutase and stress tolerance. *Annual Review of Plant Biology*, *43*, 83-116.

Bowler, C., Van Camp, W., Van Montagu, M., Inze, D. & Asada, K. (1994). Superoxide dismutase in plants. *Critical Reviews in Plant Sciences*, *13*, 199-218.

Bueno, P., Varela, J., Gimenez-Gallego, G. & del Rio, L. A. (1995). Peroxisomal copper, zinc superoxide dismutase (characterization of the isoenzyme from watermelon cotyledons). *Plant Physiology*, *108*, 1151-1160.

Carter, M. F. & Grace, J. B. (1990). Relationships between flooding tolerance, life history, and short-term competitive performance in three species of Polygonum. *American Journal of Botany*, *77*, 381-387.

Chen, T., Li, W., Hu, X., Guo, J., Liu, A. & Zhang, B. (2015). A cotton MYB transcription factor, GbMYB5, is positively involved in plant adaptive response to drought stress. *Plant and Cell Physiology*, *56*, 917-929.

Chu, C. C., Lee, W. C., Guo, W. Y., Pan, S. M., Chen, L. J., Li, H. M. & Jinn, T. L. (2005). A copper chaperone for superoxide dismutase that confers three types of copper/zinc superoxide dismutase activity in Arabidopsis. *Plant Physiology*, *139*, 425-436.

Chu, X., Wang, C., Chen, X., Lu, W., Li, H., Wang, X., Hao, L. & Guo, X. (2015). The cotton WRKY gene GhWRKY41 positively regulates salt and drought stress tolerance in transgenic *Nicotiana benthamiana*. *PloS One*, *10*, e0143022. https://doi.org/10.1371/journal.pone. 01430 22.

Cingel, A., Savić, J., Ćosić, T., Zdravković-Korać, S., Momčilović, I., Smigocki, A. & Ninković, S. (2014). Pyramiding rice cystatin OCI and OCII genes in transgenic potato (*Solanum tuberosum* L.) for resistance to Colorado potato beetle (*Leptinotarsa decemlineata* Say). *Euphytica*, *198*, 425-438.

Corpet, F. (1988). Multiple sequence alignment with hierarchical clustering. *Nucleic Acids Research*, *16*, 10881-10890.

Culotta, V. C., Yang, M. & O'Halloran, T. V. (2006). Activation of superoxide dismutases: putting the metal to the pedal. *Biochimica et Biophysica Acta (BBA)-Molecular Cell Research*, *1763*, 747-758.

del Río, L. A., Corpas, F. J., Sandalio, L. M., Palma, J. M., Gómez, M. & Barroso, J. B. (2002). Reactive oxygen species, antioxidant systems and nitric oxide in peroxisomes. *Journal of Experimental Botany*, *53*, 1255-1272.

Dhindsa, R. S. & Matowe, W. (1981). Drought tolerance in two mosses: correlated with enzymatic defence against lipid peroxidation. *Journal of Experimental Botany*, *32*, 79-91.

Diaz-Vivancos, P., Faize, M., Barba-Espin, G., Faize, L., Petri, C., Hernández, J. A. & Burgos, L. (2013). Ectopic expression of cytosolic superoxide dismutase and ascorbate peroxidase leads to salt stress tolerance in transgenic plums. *Plant Biotechnology Journal*, *11*, 976-985.

Dubos, C., Stracke, R., Grotewold, E., Weisshaar, B., Martin, C. & Lepiniec, L. (2010). MYB transcription factors in *Arabidopsis*. *Trends in Plant Science*, *15*, 573-581.

Dugas, D. V. & Bartel, B. (2008). Sucrose induction of Arabidopsis miR398 represses two Cu/Zn superoxide dismutases. *Plant Molecular Biology*, *67*, 403-417.

Dupont, C. L., Neupane, K., Shearer, J. & Palenik, B. (2008). Diversity, function and evolution of genes coding for putative Ni-containing superoxide dismutases. *Environmental Microbiology*, *10*, 1831-1843.

Else, M. A., Janowiak, F., Atkinson, C. J. & Jackson, M. B. (2009). Root signals and stomatal closure in relation to photosynthesis, chlorophyll a fluorescence and adventitious rooting of flooded tomato plants. *Annals of Botany*, *103*, 313-323.

Faize, M., Burgos, L., Faize, L., Piqueras, A., Nicolas, E., Barba-Espin, G., Clemente-Moreno, M. J., Alcobendas, R., Artlip, T. & Hernandez, J. A. (2011). Involvement of cytosolic ascorbate peroxidase and Cu/Zn-superoxide dismutase for improved tolerance against drought stress. *Journal of Experimental Botany*, *62*, 2599-2613.

Feng, W., Hongbin, W., Bing, L. & Jinfa, W. (2006). Cloning and characterization of a novel splicing isoform of the iron-superoxide dismutase gene in rice (*Oryza sativa* L.). *Plant Cell Reports*, *24*, 734.

Feng, X., Lai, Z., Lin, Y., Lai, G. & Lian, C. (2015). Genome-wide identification and characterization of the superoxide dismutase gene family in *Musa acuminata* cv. Tianbaojiao (AAA group). *BMC Genomics*, *16*(1), 823.

Feng, K., Yu, J., Cheng, Y., Ruan, M., Wang, R., Ye, Q., Zhou, G., Li, Z., Yao, Z., Yang, Y., Zheng, Q. & Zheng, Q. (2016). The SOD gene family in tomato: identification, phylogenetic relationships, and expression patterns. *Frontiers in Plant Science*, *7*, 1279. http://doi.org/10.3389/fpls.2016.01279.

Fernández-Ocaña, A., Chaki, M., Luque, F., Gómez-Rodríguez, M. V., Carreras, A., Valderrama, R. & Barroso, J. B. (2011). Functional analysis of superoxide dismutases (SODs) in sunflower under biotic and abiotic stress conditions. Identification of two new genes of mitochondrial Mn-SOD. *Journal of Plant Physiology*, *168*, 1303-1308.

Fink, R. C. & Scandalios, J. G. (2002). Molecular evolution and structure–function relationships of the superoxide dismutase gene families in angiosperms and their relationship to other eukaryotic and prokaryotic superoxide dismutases. *Archives of Biochemistry and Biophysics*, *399*, 19-36.

Flexas, J., Bota, J., Loreto, F., Cornic, G. & Sharkey, T. D. (2004). Diffusive and metabolic limitations to photosynthesis under drought and salinity in C3 plants. *Plant Biology*, *6*, 269-279.

Fridovich, I. (1986). Superoxide dismutases. *Advances in Enzymology and Related Areas of Molecular Biology*, *58*, 61-97.

Fukao, T. & Bailey-Serres, J. (2004). Plant responses to hypoxia–is survival a balancing act? *Trends in Plant Science*, *9*, 449-456.

Gechev, T., Willekens, H., Van Montagu, M., Inzé, D., Van Camp, W., Toneva, V. & Minkov, I. (2003). Different responses of tobacco antioxidant enzymes to light and chilling stress. *Journal of Plant Physiology*, *160*, 509-515.

Gomez, J. M., Hernandez, J. A., Jimenez, A., Del Rio, L. A. & Sevilla, F. (1999). Differential response of antioxidative enzymes of chloroplasts and mitochondria to long-term NaCl stress of pea plants. *Free Radical Research*, *31*(sup1), 11-18.

Guan, Q., Liao, X., He, M., Li, X., Wang, Z., Ma, H. & Liu, S. (2017). Tolerance analysis of chloroplast *OsCu/Zn-SOD* overexpressing rice under NaCl and $NaHCO_3$ stress. *PloS One*, *12*(10), e0186052. https://doi.org/10.1371/journal.pone.0186052.

Guo, H., Wang, Y., Wang, L., Hu, P., Wang, Y., Jia, Y. & Yang, C. (2017). Expression of the MYB transcription factor gene *BplMYB46* affects abiotic stress tolerance and secondary cell wall deposition in *Betula platyphylla*. *Plant Biotechnology Journal*, *15*(1), 107-121.

Gupta, A. S., Webb, R. P., Holaday, A. S. & Allen, R. D. (1993). Overexpression of superoxide dismutase protects plants from oxidative stress (induction of ascorbate peroxidase in superoxide dismutase-overexpressing plants). *Plant Physiology*, *103*(4), 1067-1073.

Halpin, C., Barakate, A., Askari, B. M., Abbott, J. C. & Ryan, M. D. (2001). Enabling technologies for manipulating multiple genes on complex pathways. In N. C. Carpita, M. Campbell, & M. Tierney (Eds.), *Plant Cell Walls*, (pp. 295-310). Dordrecht, Netherlands: Springer.

Halpin, C. (2005). Gene stacking in transgenic plants-the challenge for 21st century plant biotechnology. *Plant Biotechnology Journal*, *3*, 141-155.

Horton, R. F. (1992). Submergence-promoted growth of petioles of *Ranunculus pygmaeus* Wahl. *Aquatic Botany*, *44*, 23-30.

Hu, W. H., Song, X. S., Shi, K., Xia, X. J., Zhou, Y. H. & Yu, J. Q. (2008). Changes in electron transport, superoxide dismutase and ascorbate peroxidase isoenzymes in chloroplasts and mitochondria of cucumber leaves as influenced by chilling. *Photosynthetica*, *46*, 581-588.

Huang, J., Sun, S. J., Xu, D. Q., Yang, X., Bao, Y. M., Wang, Z. F. & Zhang, H. (2009). Increased tolerance of rice to cold, drought and oxidative stresses mediated by the overexpression of a gene that encodes the zinc finger protein ZFP245. *Biochemical and Biophysical Research Communications*, *389*, 556-561.

Huang, X. S., Liu, J. H. & Chen, X. J. (2010). Overexpression of PtrABF gene, a bZIP transcription factor isolated from *Poncirus trifoliata*, enhances dehydration and drought tolerance in tobacco via scavenging

ROS and modulating expression of stress-responsive genes. *BMC Plant Biology*, *10*, 230.

Irigoyen, J. J., De Juan, J. P. & Sanchez-Diaz, M. (1996). Drought enhances chilling tolerance in a chilling-sensitive maize (*Zea mays*) variety. *New Phytologist*, *134*, 53-59.

Jacobsen, S. E., Jensen, C. R. & Liu, F. (2012). Improving crop production in the arid Mediterranean climate. *Field Crops Research*, *128*, 34-47.

Jarvis, P. (2008). Targeting of nucleus-encoded proteins to chloroplasts in plants. *New Phytologist*, *179*, 257-285.

Jing, X., Hou, P., Lu, Y., Deng, S., Li, N., Zhao, R. & Chen, S. (2015). Overexpression of copper/zinc superoxide dismutase from mangrove *Kandelia candel* in tobacco enhances salinity tolerance by the reduction of reactive oxygen species in chloroplast. *Frontiers in Plant Science*, *6*, 23. http://doi.org/10.3389/fpls.2015.00023.

Kanematsu, S. & Asada, K. (1979). Ferric and manganic superoxide dismutases in *Euglena gracilis*. *Archives of Biochemistry and Biophysics*, *195*, 535-545.

Kanematsu, S. & Asada, K. (1989). CuZn-superoxide dismutases in rice: occurrence of an active, monomeric enzyme and two types of isozyme in leaf and non-photosynthetic tissues. *Plant and Cell Physiology*, *30*, 381-391.

Kanematsu, S. & Asada, K. (1990). Characteristic amino acid sequences of chloroplast and cytosol isozymes of CuZn-superoxide dismutase in spinach, rice and horsetail. *Plant and Cell Physiology*, *31*, 99-112.

Kanematsu, S., Okayasu, M. & Ueno, S. (2013). Atypical cytosol-localized Fe-superoxide dismutase in the moss *Pogonatum inflexum*. *Bull. Minamikyushu Univ.*, *43A*, *23*, 31.

Karpinska, B., Karlsson, M., Schinkel, H., Streller, S., Süss, K. H., Melzer, M. & Wingsle, G. (2001). A novel superoxide dismutase with a high isoelectric point in higher plants. Expression, regulation, and protein localization. *Plant Physiology*, *126*, 1668-1677.

Katyshev, A. I., Rogozin, I. B. & Konstantinov, Y. M. (2006a) Identification of new superoxide dismutase transcripts in plants by est analysis: Alternative polyadenylation and splicing events. In

Proceedings of the Fifth International Conference on Bioinformatics of Genome Regulation and Structure. Novosibirs: Russian Acad Sci Siberian Branch, (pp. 61-64).

Katyshev, A. I., Konstantinov, Y. M., and Kobzev, V. F. (2006b) Characterization of Mn and Cu/Zn containing superoxide dismutase gene transcripts in *Larix gmelinii*. *Molecular Biology*, *40*, 327-329.

Kennedy, R. A., Rumpho, M. E. & Fox, T. C. (1992). Anaerobic metabolism in plants. *Plant Physiology*, *100*, 1-6.

Kim, H. J. & Triplett, B. (2008). Involvement of extracellular Cu/Zn superoxide dismutase in cotton fiber primary and secondary cell wall biosynthesis. *Plant Signaling & Behavior*, *3*, 1119-1121.

Kliebenstein, D. J., Monde, R. A. & Last, R. L. (1998). Superoxide dismutase in Arabidopsis: an eclectic enzyme family with disparate regulation and protein localization. *Plant Physiology*, *118*, 637-650.

Kumar, R. (2014). Role of microRNAs in biotic and abiotic stress responses in crop plants. *Applied Biochemistry and Biotechnology*, *174*, 93-115.

Kumutha, D., Ezhilmathi, K., Sairam, R. K., Srivastava, G. C., Deshmukh, P. S. & Meena, R. C. (2009). Waterlogging induced oxidative stress and antioxidant activity in pigeon pea genotypes. *Biologia Plantarum*, *53*, 75-84.

Kwon, S. Y., Jeong, Y. J., Lee, H. S., Kim, J. S., Cho, K. Y., Allen, R. D. & Kwak, S. S. (2002). Enhanced tolerances of transgenic tobacco plants expressing both superoxide dismutase and ascorbate peroxidase in chloroplasts against methyl viologen-mediated oxidative stress. *Plant, Cell & Environment*, *25*, 873-882.

Lee, D. H., Kim, Y. S. & Lee, C. B. (2001). The inductive responses of the antioxidant enzymes by salt stress in the rice (Oryza sativa L.). *Journal of Plant Physiology*, *158*(6), 737-745.

Lee, Y. P., Kim, S. H., Bang, J. W., Lee, H. S., Kwak, S. S. & Kwon, S. Y. (2007). Enhanced tolerance to oxidative stress in transgenic tobacco plants expressing three antioxidant enzymes in chloroplasts. *Plant Cell Reports*, *26*, 591-598.

Li, C., Han, L. B. & Zhang, X. (2012). Enhanced drought tolerance of tobacco overexpressing *OJERF* gene is associated with alteration in proline and antioxidant metabolism. *Journal of the American Society for Horticultural Science*, *137*, 107-113.

Li, C. R., Liang, D. D., Li, J., Duan, Y. B., Li, H. A. O., Yang, Y. C., Qin, R. Y., Wei, P. C. & Yang, J. B. (2013). Unravelling mitochondrial retrograde regulation in the abiotic stress induction of rice Alternative Oxidase 1 genes. *Plant, Cell & Environment*, *36*, 775-788.

Li, J. B., Luan, Y. S. & Liu, Z. (2015). Overexpression of *SpWRKY1* promotes resistance to *Phytophthora nicotianae* and tolerance to salt and drought stress in transgenic tobacco. *Physiologia Plantarum*, *155*, 248-266.

Liang, C., Meng, Z., Meng, Z., Malik, W., Yan, R., Lwin, K. M. & Zhu, T. (2016). GhABF2, a bZIP transcription factor, confers drought and salinity tolerance in cotton (*Gossypium hirsutum* L.). *Scientific Reports*, *6*, 35040.

Lim, S., Kim, Y. H., Kim, S. H., Kwon, S. Y., Lee, H. S., Kim, J. S. & Kwak, S. S. (2007). Enhanced tolerance of transgenic sweet potato plants that express both CuZnSOD and APX in chloroplasts to methyl viologen-mediated oxidative stress and chilling. *Molecular Breeding*, *19*, 227-239.

Lin, Y. L. & Lai, Z. X. (2013). Superoxide dismutase multigene family in longan somatic embryos: a comparison of CuZn-SOD, Fe-SOD, and Mn-SOD gene structure, splicing, phylogeny, and expression. *Molecular Breeding*, *32*, 595-615.

Liu, W., Yu, K., He, T., Li, F., Zhang, D. & Liu, J. (2013). The low temperature induced physiological responses of *Avena nuda* L., a cold-tolerant plant species. *The Scientific World Journal.* Volume 2013, Article ID 658793, 7 pages. http://dx.doi.org/10.1155/2013/658793.

Liu, C., Wang, Y., Pan, K., Li, W., Zhang, L., Shen, X., Liu, L. & Deng, M. (2014). Responses of the antioxidant defense system to drought stress in the leaves of *Fargesia denudata* seedlings, the staple food of the giant panda. *Russian Journal of Plant Physiology*, *61*, 374-383.

Liu, Z. B., Zhang, W. J., Gong, X. D., Zhang, Q. & Zhou, L. R. (2015). A Cu/Zn superoxide dismutase from *Jatropha curcas* enhances salt tolerance of *Arabidopsis thaliana. Genetics and Molecular Research, 14*, 2086-2098.

Logan, B. A., Kornyeyev, D., Hardison, J. & Holaday, A. S. (2006). The role of antioxidant enzymes in photo protection. *Photosynthesis Research, 88*, 119-132.

Luo, X., Wu, J., Li, Y., Nan, Z., Guo, X., Wang, Y., Zhang, A., Wang, Z., Xia, G. & Tian, Y. (2013). Synergistic effects of *GhSOD1* and *GhCAT1* overexpression in cotton chloroplasts on enhancing tolerance to methyl viologen and salt stresses. *PLoS ONE, 8*(1), e54002. http://doi.org/10.1371/journal.pone.0054002.

Lutz, C. S. (2008). Alternative polyadenylation: a twist on mRNA 3′ end formation. *ACS Chemical Biology, 3*(10), 609-617.

Ma, C., Burd, S. & Lers, A. (2015). miR408 is involved in abiotic stress responses in Arabidopsis. *The Plant Journal, 84*, 169-187.

Malan, C., Greyling, M. M. & Gressel, J. (1990). Correlation between CuZn superoxide dismutase and glutathione reductase, and environmental and xenobiotic stress tolerance in maize inbreds. *Plant Science, 69*, 157-166.

McKersie, B. D., Bowley, S. R., Harjanto, E. & Leprince, O. (1996). Water-deficit tolerance and field performance of transgenic alfalfa overexpressing superoxide dismutase. *Plant Physiology, 111*, 1177-1181.

McKersie, B. D., Murnaghan, J., Jones, K. S. & Bowley, S. R. (2000). Iron-superoxide dismutase expression in transgenic alfalfa increases winter survival without a detectable increase in photosynthetic oxidative stress tolerance. *Plant Physiology, 122*, 1427-1438.

Meier, B., Sehn, A. P., Schininà, M. E., & Barra, D. (1994). *In vivo* incorporation of copper into the iron-exchangeable and manganese-exchangeable superoxide dismutase from *Propionibacterium shermanii. The FEBS Journal, 219*(1-2), 463-468.

Miller, G., Suzuki, N., Ciftci-Yilmaz, S. & Mittler, R. O. N. (2010). Reactive oxygen species homeostasis and signalling during drought and salinity stresses. *Plant, Cell & Environment*, *33*, 453-467.

Møller, I. M. (2001). Plant mitochondria and oxidative stress: electron transport, NADPH turnover, and metabolism of reactive oxygen species. *Annual Review of Plant Biology*, *52*, 561-591.

Momčilović, I., Pantelić, D., Hfidan, M., Savić, J. & Vinterhalter, D. (2014). Improved procedure for detection of superoxide dismutase isoforms in potato, *Solanum tuberosum* L. *Acta Physiologiae Plantarum*, *36*, 2059-2066.

Moran, J. F., James, E. K., Rubio, M. C., Sarath, G., Klucas, R. V. & Becana, M. (2003). Functional characterization and expression of a cytosolic iron-superoxide dismutase from cowpea root nodules. *Plant Physiology*, *133*, 773-782.

Muñoz, I. G., Moran, J. F., Becana, M. & Montoya, G. (2005). The crystal structure of an eukaryotic iron superoxide dismutase suggests intersubunit cooperation during catalysis. *Protein Science*, *14*, 387-394.

Myouga, F., Hosoda, C., Umezawa, T., Iizumi, H., Kuromori, T., Motohashi, R., Shono, Y., Nagata, N., Ikeuchi, M. & Shinozaki, K. (2008). A heterocomplex of iron superoxide dismutases defends chloroplast nucleoids against oxidative stress and is essential for chloroplast development in *Arabidopsis*. *The Plant Cell*, *20*, 3148-3162.

Nagae, M., Nakata, M. & Takahashi, Y. (2008). Identification of negative cis-acting elements in response to copper in the chloroplastic iron superoxide dismutase gene of the moss *Barbula unguiculata*. *Plant Physiology*, *146*, 1687-1696.

Nath, K., Kumar, S., Poudyal, R. S., Yang, Y. N., Timilsina, R., Park, Y. S., Puneet, J. N., Chauhan, P. S., Pant, B. & Lee, C. H. (2014). Developmental stage-dependent differential gene expression of superoxide dismutase isoenzymes and their localization and physical interaction network in rice (*Oryza sativa* L.). *Genes & Genomics*, *36*, 45-55.

Negi, N. P., Shrivastava, D. C., Sharma, V. & Sarin, N. B. (2015). Overexpression of CuZnSOD from *Arachis hypogaea* alleviates salinity and drought stress in tobacco. *Plant Cell Reports*, *34*, 1109-1126.

Negi, N. P., Shrivastava, D., Shekhar, S., Sharma, V. & Sarin, N. B. (2016). Simultaneous overexpression of CuZnSOD and cAPX from *Arachis hypogaea* leads to salinity stress tolerance in tobacco. *In Vitro Cellular & Developmental Biology-Plant*, *52*, 484-491.

Ogawa, K. I., Kanematsu, S., Takabe, K. & Asada, K. (1995). Attachment of CuZn-superoxide dismutase to thylakoid membranes at the site of superoxide generation (PSI) in spinach chloroplasts: detection by immuno-gold labeling after rapid freezing and substitution method. *Plant and Cell Physiology*, *36*, 565-573.

Ogawa, K. I., Kanematsu, S. & Asada, K. (1996). Intra-and extra-cellular localization of "cytosolic" CuZn-superoxide dismutase in spinach leaf and hypocotyl. *Plant and Cell Physiology*, *37*, 790-799.

Okada, S., Kanematsu, S. & Asada, K. (1979). Intracellular distribution of manganese and ferric superoxide dismutases in blue-green algae. *FEBS Letters*, *103*, 106-110.

Pal, A. K., Acharya, K., Vats, S. K., Kumar, S. & Ahuja, P. S. (2013). Over-expression of PaSOD in transgenic potato enhances photosynthetic performance under drought. *Biologia Plantarum*, *57*, 359-364.

Pan, Y., Wu, L. J. & Yu, Z. L. (2006). Effect of salt and drought stress on antioxidant enzymes activities and SOD isoenzymes of liquorice (*Glycyrrhiza uralensis* Fisch). *Plant Growth Regulation*, *49*, 157-165.

Payton, P., Allen, R. D., Trolinder, N. & Holaday, A. S. (1997). Over-expression of chloroplast-targeted Mn superoxide dismutase in cotton (*Gossypium hirsutum* L., cv. Coker 312) does not alter the reduction of photosynthesis after short exposures to low temperature and high light intensity. *Photosynthesis Research*, *52*, 233-244.

Perl, A., Perl-Treves, R., Galili, S., Aviv, D., Shalgi, E., Malkin, S. & Galun, E. (1993). Enhanced oxidative-stress defense in transgenic potato expressing tomato Cu, Zn superoxide dismutases. *TAG Theoretical and Applied Genetics*, *85*, 568-576.

Perl-Treves, R. & Galun, E. (1991). The tomato Cu, Zn superoxide dismutase genes are developmentally regulated and respond to light and stress. *Plant Molecular Biology*, *17*, 745-760.

Pilon, M., Ravet, K. & Tapken, W. (2011). The biogenesis and physiological function of chloroplast superoxide dismutases. *Biochimica et Biophysica Acta (BBA)-Bioenergetics*, *1807*, 989-998.

Pitcher, L. H., Brennan, E., Hurley, A., Dunsmuir, P., Tepperman, J. M. & Zilinskas, B. A. (1991). Overproduction of petunia chloroplastic copper/zinc superoxide dismutase does not confer ozone tolerance in transgenic tobacco. *Plant Physiology*, *97*, 452-455.

Prashanth, S. R., Sadhasivam, V. & Parida, A. (2008). Over expression of cytosolic copper/zinc superoxide dismutase from a mangrove plant *Avicennia marina* in indica rice var Pusa Basmati-1 confers abiotic stress tolerance. *Transgenic Research*, *17*, 281-291.

Prochazkova, D., Sairam, R. K., Srivastava, G. C. & Singh, D. V. (2001). Oxidative stress and antioxidant activity as the basis of senescence in maize leaves. *Plant Science*, *161*, 765-771.

Roach, T. & Krieger-Liszkay, A. (2014). Regulation of photosynthetic electron transport and photo inhibition. *Current Protein and Peptide Science*, *15*, 351-362.

Robert, X. & Gouet, P. (2014). Deciphering key features in protein structures with the new ENDscript server. *Nucleic Acids Research*, *42*(W1), W320-W324.

Rose, A. S. & Hildebrand, P. W. (2015). NGL Viewer: a web application for molecular visualization. *Nucleic Acids Research*, *43*(W1), W576-W579.

Rubio, M. C., Becana, M., Sato, S., James, E. K., Tabata, S. & Spaink, H. P. (2007). Characterization of genomic clones and expression analysis of the three types of superoxide dismutases during nodule development in *Lotus japonicus*. *Molecular Plant-microbe interactions*, *20*, 262-275.

Samis, K., Bowley, S. & McKersie, B. (2002). Pyramiding Mn-superoxide dismutase transgenes to improve persistence and biomass production in alfalfa. *Journal of Experimental Botany*, *53*, 1343-1350.

Schinkel, H., Hertzberg, M. & Wingsle, G. (2001). A small family of novel CuZn-superoxide dismutases with high isoelectric points in hybrid aspen. *Planta*, *213*, 272-279.

Shafi, A., Chauhan, R., Gill, T., Swarnkar, M. K., Sreenivasulu, Y., Kumar, S., Kumar, N., Shankar, R., Ahuja, P. S. & Singh, A. K. (2015). Expression of SOD and APX genes positively regulates secondary cell wall biosynthesis and promotes plant growth and yield in Arabidopsis under salt stress. *Plant Molecular Biology*, *87*, 615-631.

Shafi, A., Pal, A. K., Sharma, V., Kalia, S., Kumar, S., Ahuja, P. S. & Singh, A. K. (2017). Transgenic Potato Plants Overexpressing SOD and APX Exhibit Enhanced Lignification and Starch Biosynthesis with Improved Salt Stress Tolerance. *Plant Molecular Biology Reporter*, *35*, 504-518.

Sharma, P. & Dubey, R. S. (2005). Drought induces oxidative stress and enhances the activities of antioxidant enzymes in growing rice seedlings. *Plant Growth Regulation*, *46*, 209-221.

Shi, W., Liu, D., Hao, L., Wu, C. A., Guo, X. & Li, H. (2014). *GhWRKY39*, a member of the WRKY transcription factor family in cotton, has a positive role in disease resistance and salt stress tolerance. *Plant Cell, Tissue and Organ Culture*, *118*, 17-32.

Shingote, P. R., Kawar, P. G., Pagariya, M. C., Kuhikar, R. S., Thorat, A. S. & Babu, K. H. (2015). SoMYB18, a sugarcane MYB transcription factor improves salt and dehydration tolerance in tobacco. *Acta Physiologiae Plantarum*, *37*, 1-12.

Shiraya, T., Mori, T., Maruyama, T., Sasaki, M., Takamatsu, T., Oikawa, K., Itoh, K., Kaneko, K., Ichikawa, H. & Mitsui, T. (2015). Golgi/plastid-type manganese superoxide dismutase involved in heat-stress tolerance during grain filling of rice. *Plant Biotechnology Journal*, *13*, 1251-1263.

Sofo, A., Dichio, B., Xiloyannis, C. & Masia, A. (2005). Antioxidant defences in olive trees during drought stress: changes in activity of some antioxidant enzymes. *Functional Plant Biology*, *32*, 45-53.

Srivastava, V., Srivastava, M. K., Chibani, K., Nilsson, R., Rouhier, N., Melzer, M. & Wingsle, G. (2009). Alternative splicing studies of the reactive oxygen species gene network in *Populus* reveal two isoforms of high-isoelectric-point superoxide dismutase. *Plant Physiology*, *149*, 1848-1859.

Sun, S. J., Guo, S. Q., Yang, X., Bao, Y. M., Tang, H. J., Sun, H. & Zhang, H. S. (2010). Functional analysis of a novel Cys2/His2-type zinc finger protein involved in salt tolerance in rice. *Journal of Experimental Botany*, *61*, 2807-2818.

Sun, J., Hu, W., Zhou, R., Wang, L., Wang, X., Wang, Q. & Yang, G. (2015). The *Brachypodium distachyon BdWRKY36* gene confers tolerance to drought stress in transgenic tobacco plants. *Plant Cell Reports*, *34*, 23-35.

Sundaram, S., Khanna, S. & Khanna-Chopra, R. (2009). Purification and characterization of thermostable monomeric chloroplastic Cu/Zn superoxide dismutase from *Chenopodium murale*. *Physiology and Molecular Biology of Plants*, *15*, 199-209.

Sunkar, R. & Zhu, J. K. (2004). Novel and stress-regulated microRNAs and other small RNAs from Arabidopsis. *The Plant Cell*, *16*, 2001-2019.

Sunkar, R., Kapoor, A. & Zhu, J. K. (2006). Posttranscriptional induction of two Cu/Zn superoxide dismutase genes in Arabidopsis is mediated by downregulation of miR398 and important for oxidative stress tolerance. *The Plant Cell*, *18*, 2051-2065.

Szal, B., Drozd, M. & Rychter, A. M. (2004). Factors affecting determination of superoxide anion generated by mitochondria from barley roots after anaerobiosis. *Journal of Plant Physiology*, *161*, 1339-1346.

Tanaka, K., Takio, S., Yamamoto, I. & Satoh, T. (1996). Purification of the cytosolic CuZn-superoxide dismutase (CuZn-SOD) of *Marachantia paleacea* var. diptera and its resemblance to CuZn-SOD from chloroplasts. *Plant and Cell Physiology*, *37*, 523-529.

Tang, W., Charles, T. M. & Newton, R. J. (2005). Overexpression of the pepper transcription factor CaPF1 in transgenic *Virginia pine* (*Pinus virginiana* Mill.) confers multiple stress tolerance and enhances organ growth. *Plant Molecular Biology*, *59*, 603-617.

Tepperman, J. M. & Dunsmuir, P. (1990). Transformed plants with elevated levels of chloroplastic SOD are not more resistant to superoxide toxicity. *Plant Molecular Biology*, *14*, 501-511.

Tseng, M. J., Liu, C. W. & Yiu, J. C. (2007). Enhanced tolerance to sulfur dioxide and salt stress of transgenic Chinese cabbage plants expressing both superoxide dismutase and catalase in chloroplasts. *Plant Physiology and Biochemistry*, *45*, 822-833.

Ullah, A., Sun, H., Hakim, Yang, X. & Zhang, X. (2017). A novel cotton WRKY-gene, *GhWRKY6-like*, improves salt tolerance by activating the ABA signalling pathway and scavenging of reactive oxygen species. *Physiologia Plantarum*. Accepted Author Manuscript. http://doi.org/10.1111/ppl.12651.

Ushimaru, T., Kanematsu, S., Shibasaka, M. & Tsuji, H. (1999). Effect of hypoxia on the antioxidative enzymes in aerobically grown rice (*Oryza sativa*) seedlings. *Physiologia Plantarum*, *107*, 181-187.

Van Breusegem, F., Slooten, L., Stassart, J. M., Botterman, J., Moens, T., Van Montagu, M. & Inzé, D. (1999a). Effects of overproduction of tobacco MnSOD in maize chloroplasts on foliar tolerance to cold and oxidative stress. *Journal of Experimental Botany*, *50*, 71-78.

Van Breusegem, F., Slooten, L., Stassart, J. M., Moens, T., Botterman, J., Van Montagu, M. & Inzé, D. (1999b). Overproduction of *Arabidopsis thaliana* FeSOD confers oxidative stress tolerance to transgenic maize. *Plant and Cell Physiology*, *40*, 515-523.

Van Camp, W., Capiau, K., Van Montagu, M., Inze, D. & Slooten, L. (1996). Enhancement of oxidative stress tolerance in transgenic tobacco plants overproducing Fe-superoxide dismutase in chloroplasts. *Plant Physiology*, *112*, 1703-1714.

Van Der Straeten, D., Zhou, Z., Prinsen, E., Van Onckelen, H. A. & Van Montagu, M. C. (2001). A comparative molecular-physiological study of submergence response in lowland and deep water rice. *Plant Physiology*, *125*, 955-968.

Vyas, D. & Kumar, S. (2005). Purification and partial characterization of a low temperature responsive Mn-SOD from tea (*Camellia sinensis* (L.) O. Kuntze). *Biochemical and Biophysical Research Communications*, *329*, 831-838.

Wang, Y., Ying, Y., Chen, J. & Wang, X. (2004). Transgenic Arabidopsis overexpressing Mn-SOD enhanced salt-tolerance. *Plant Science*, *167*(4), 671-677.

Wang, F. Z., Wang, Q. B., Kwon, S. Y., Kwak, S. S. & Su, W. A. (2005). Enhanced drought tolerance of transgenic rice plants expressing a pea manganese superoxide dismutase. *Journal of Plant Physiology*, *162*, 465-472.

Wang, Y. C., Qu, G. Z., Li, H. Y., Wu, Y. J., Wang, C., Liu, G. F. & Yang, C. P. (2010a). Enhanced salt tolerance of transgenic poplar plants expressing a manganese superoxide dismutase from *Tamarix androssowii*. *Molecular Biology Reports*, *37*, 1119.

Wang, Y., Gao, C., Liang, Y., Wang, C., Yang, C. & Liu, G. (2010b). A novel bZIP gene from *Tamarix hispida* mediates physiological responses to salt stress in tobacco plants. *Journal of Plant Physiology*, *167*, 222-230.

Wang, X., Zeng, J., Li, Y., Rong, X., Sun, J., Sun, T. & He, G. (2015). Expression of *TaWRKY44*, a wheat *WRKY* gene, in transgenic tobacco confers multiple abiotic stress tolerances. *Frontiers in Plant Science*, *6*, 615. http://doi.org/10.3389/fpls.2015.00615.

Wang, F., Liu, J., Zhou, L., Pan, G., Li, Z. & Cheng, F. (2016a). Senescence-specific change in ROS scavenging enzyme activities and regulation of various SOD isozymes to ROS levels in psf mutant rice leaves. *Plant Physiology and Biochemistry*, *109*, 248-261.

Wang, W., Xia, M. X., Chen, J., Yuan, R., Deng, F. N. & Shen, F. F. (2016b). Gene expression characteristics and regulation mechanisms of superoxide dismutase and its physiological roles in plants under stress. *Biochemistry (Moscow)*, *81*, 465-480.

Wang, W., Zhang, X., Deng, F., Yuan, R. & Shen, F. (2017). Genome-wide characterization and expression analyses of superoxide dismutase (SOD) genes in *Gossypium hirsutum*. *BMC Genomics*, *18*, 376.

Wei, Q., Luo, Q., Wang, R., Zhang, F., He, Y., Zhang, Y. & He, G. (2017). A Wheat R2R3-type MYB Transcription Factor TaODORANT1 Positively Regulates Drought and Salt Stress Responses in Transgenic Tobacco Plants. *Frontiers in Plant Science*, *8*, 1374.

Wu, L., Zhang, Z., Zhang, H., Wang, X. C. & Huang, R. (2008). Transcriptional modulation of ethylene response factor protein JERF3 in the oxidative stress response enhances tolerance of tobacco seedlings to salt, drought, and freezing. *Plant Physiology*, *148*, 1953-1963.

Xu, J., Yang, J., Duan, X., Jiang, Y. & Zhang, P. (2014). Increased expression of native cytosolic Cu/Zn superoxide dismutase and ascorbate peroxidase improves tolerance to oxidative and chilling stresses in cassava (*Manihot esculenta* Crantz). *BMC Plant Biology*, *14*, 208.

Xu, S. C., Li, Y. P., Jin, H. U., Guan, Y. J., Zheng, Y. Y. & Zhu, S. J. (2010). Responses of antioxidant enzymes to chilling stress in tobacco seedlings. *Agricultural Sciences in China*, *9*, 1594-1601.

Yamasaki, H., Hayashi, M., Fukazawa, M., Kobayashi, Y. & Shikanai, T. (2009). SQUAMOSA promoter binding protein–like7 is a central regulator for copper homeostasis in *Arabidopsis*. *The Plant Cell*, *21*, 347-361.

Yan, B., Dai, Q., Liu, X., Huang, S. & Wang, Z. (1996). Flooding-induced membrane damage, lipid oxidation and activated oxygen generation in corn leaves. *Plant and Soil*, *179*, 261-268.

Yang, A., Dai, X. & Zhang, W. H. (2012). A R2R3-type MYB gene, OsMYB2, is involved in salt, cold, and dehydration tolerance in rice. *Journal of Experimental Botany*, *63*, 2541-2556.

Zang, D., Li, H., Xu, H., Zhang, W., Zhang, Y., Shi, X. & Wang, Y. (2016). An Arabidopsis zinc finger protein increases abiotic stress tolerance by regulating sodium and potassium homeostasis, reactive oxygen species scavenging and osmotic potential. *Frontiers in Plant Science*, *7*, 1272. http://doi.org/10.3389/fpls.2016.01272.

Zang, D., Wang, C., Ji, X. & Wang, Y. (2015). *Tamarix hispida* zinc finger protein ThZFP1 participates in salt and osmotic stress tolerance by increasing proline content and SOD and POD activities. *Plant Science*, *235*, 111-121.

Zhang, D. Y., Yang, H. L., Li, X. S., Li, H. Y. & Wang, Y. C. (2014). Overexpression of *Tamarix albiflonum TaMnSOD* increases drought tolerance in transgenic cotton. *Molecular Breeding*, *34*, 1-11.

Zhang, L., Sun, L., Zhang, L., Qiu, H., Liu, C., Wang, A. & Zhu, J. (2017). A Cu/Zn superoxide dismutase gene from *Saussurea involucrata* Kar. et Kir., *SiCSD*, enhances drought, cold and oxidative stress in transgenic tobacco. *Canadian Journal of Plant Science*, *97*, 816-826.

Zhang, W. B., Li, N. N., Qi, F., Chen, X., Hao, J., Yu, C. R. & Lin, X. F. (2014). Isolation and expression analysis of Cu/Zn superoxide dismutase genes from three *Caragana* species. *Russian Journal of Plant Physiology*, *61*, 656-663.

Zhou, T., Yang, X., Wang, L., Xu, J. & Zhang, X. (2014). *GhTZF1* regulates drought stress responses and delays leaf senescence by inhibiting reactive oxygen species accumulation in transgenic *Arabidopsis*. *Plant Molecular Biology*, *85*, 163-177.

Zhou, W., Zhao, D. & Lin, X. (1997). Effects of waterlogging on nitrogen accumulation and alleviation of waterlogging damage by application of nitrogen fertilizer and mixtalol in winter rape (*Brassica napus* L.). *Journal of Plant Growth Regulation*, *16*, 47-53.

Zhou, Y. H., Yu, J. Q., Mao, W. H., Huang, L. F., Song, X. S. & Nogués, S. (2006). Genotypic variation of Rubisco expression, photosynthetic electron flow and antioxidant metabolism in the chloroplasts of chill-exposed cucumber plants. *Plant and Cell Physiology*, *47*, 192-199.

Zhou, Y., Hu, L., Wu, H., Jiang, L. & Liu, S. (2017). Genome-wide identification and transcriptional expression analysis of cucumber superoxide dismutase (SOD) family in response to various abiotic stresses. *International Journal of Genomics*., Volume 2017, Article ID 7243973, 14 pages. https://doi.org/10.1155/2017/7243973.

Zou, M., Yuan, L., Zhu, S., Liu, S., Ge, J. & Wang, C. (2017). Effects of heat stress on photosynthetic characteristics and chloroplast ultrastructure of a heat-sensitive and heat-tolerant cultivar of wucai (*Brassica campestris* L.). *Acta Physiologiae Plantarum*, *39*, 30.

In: Superoxide Dismutase
Editor: Sergei Magliozzi

ISBN: 978-1-53613-080-5

Chapter 3

SUPEROXIDE DISMUTASE (SOD) AND ABIOTIC STRESS TOLERANCE IN PLANTS: AN OVERVIEW

Walid Saibi and Faiçal Brini*
Biotechnology and Plant Improvement Laboratory,
Centre of Biotechnology of Sfax,
University of Sfax, Sfax, Tunisia

ABSTRACT

Oxidative stress is induced by a wide range of abiotic stresses. Drought and salt stresses give usually birth to the generation of reacting oxygen species (ROS) such as hydrogen peroxide (H_2O_2) and superoxide ($O_2^{\bullet-}$). Those two species produced in a number of cellular reactions meaning various oxidoreductases such as lipoxygenases, peroxidases, NADPH oxidase and xanthine oxidase. To control the level of ROS under stress conditions, plants have involved a large scale of enzymatic and non-enzymatic antioxidants to scavenge ROS and to restore the redox homeostasis of the cells. Indeed, the superoxide dismutases (SODs) are very important attendants of the antioxidant biocatalysts due to the dismutation of O^{2-} to H_2O_2 and are located in the main cell compartments.

* Corresponding Author Email: faical.brini@cbs.rnrt.tn.

In addition, Fe-SOD, Mn-SOD and Cu/Zn-SOD are the groups according to the heavy metal co-factor used by the enzyme and they often act in a contradictory manner after abiotic stress. In this chapter, we review the oxidative stress and damage to plants. Furthermore, we try to summarize the characteristics of SOD enzymes and discuss their involvement in the tolerance of plants against abiotic stress.

Keywords: abiotic stress, oxidative stress, ROS, SOD, tolerant plants

1. Introduction

Abiotic stresses impact growth, development, and productivity, and significantly limit the global agricultural yields mainly by impairing cellular physiology and biochemistry via the increase of the reactive oxygen species (ROS) generation. Indeed, if not metabolized, ROS exceeds the status of antioxidants and cause damage to DNA, proteins, lipids, and other macromolecules, and finally cellular metabolism arrest. ROS is considered as unavoidable chemical entity of aerobic life (Halliwel and Gutteridge, 1989; Halliwell, 2006). They were considered as an aerobic metabolism by-product and whose production is confined to cellular compartments with strong electron flow. These facts include chloroplast, mitochondria and peroxisomes. ROSs are composed in part of hydrogen peroxide (H_2O_2), superoxide radical ($O_2^{\bullet -}$), hydroxyl radical ($OH^{\bullet}$) and singlet oxygen ($^1O_2$) etc. Superoxide ($O_2^{\bullet -}$) is the first species to be generated after the reduction of molecular O_2 and is considered to have strong reactivity and oxidizing capabilities. It arises in several cell compartments but the main superoxide producers are the photosynthetic electron transport chain localized in the chloroplast and the mitochondrial electron transport (respiration). Moreover, cell-wall-bound peroxidases and NADPH oxidase found in the plasma membrane and can also generate $O_2^{\bullet -}$ radicals (Arora et al., 2002; Mittler, 2002). We note also that the Hydrogen peroxide (H_2O_2) is not a radical; it has longer half-time (1 ms) and it can be generated from superoxide by dismutation, mainly in the peroxisome and the apoplast. Indeed, the described reaction may take place spontaneously at low pH or is catalyzed enzymatically via superoxide dismutase, (SOD); Arora et al., 2002). Due to its diffusibility, it can reach numerous biomolecules and affect the activity of the thiol groups oxidizing proteins. Both the hydrogen peroxide and the superoxide radicals can also be participants in the Fenton- and Haber-Weiss reactions producing the highly reactive hydroxyl radical ($^{\bullet}OH$). The last one is considered to be the most reactive oxidant with a very short half-life time (less than 1 μs) in the plant cells and can be easily formed due to the above-mentioned Fenton- and Haber-Weiss reactions. The hydroxyl radical ($^{\bullet}OH$) destructive effect includes the lipid peroxidation (LP) and the impairment of biomolecules like DNA or proteins (Dat et al., 2000; Arora et al., 2002; Gill and Tuteja, 2010). Plants, like other living organisms, have evolved a large scale of chemical components and strategies to take control over the physiological ROS production or the increased level of ROS under stress conditions like drought, salinity, UV-B radiation or heavy metal contamination (Yousuf et al., 2012). Thoses molecules, known as antioxidants, can be found in almost all cell organelles and work in a well-organized system (Gill and Tuteja, 2010). Antioxidants, which are functionally divided into two groups, namely non-enzymatic and enzymatic components, are the participants in the ROS scavenging pathways, such as the water water cycle in the chloroplast, the Halliwell-Asada (or ascorbate_glutathione) cycle in the chloroplast, mitochondria, peroxisomes, apoplast and the cytosol (Inzé and Van Montagu, 1995; Arora et al., 2002; Mittler, 2002; Zagorchev et al., 2013), and the peroxisomal glutathione peroxidase cycle, as well (Mittler, 2002). Major non-enzymatic antioxidants include ascorbic acid (AsA), glutathione (GSH), phenolic compounds, alkaloids, non-protein amino acids, and α-tocopherols. On the other hand, the battery of enzymatic antioxidants includes superoxide dismutase (SOD), catalase (CAT), ascorbate peroxidase (APX), glutathione reductase (GR), monodehydroascorbate reductase (MDHAR), dehydroascorbate reductase (DHAR), glutathione peroxidase (GPX), guaicol peroxidase (GOPX), peroxidase (POX), glutathione-S-transferase (GST) (Hasanuzzaman et al., 2012). Considering the major enzymatic antioxidants, SOD constitutes the first line of defense against abiotic stress-accrued enhanced ROS. SODs catalyze the dismutation of $O_2^{\bullet -}$ to H_2O_2 and O_2 in all subcellular compartments such as chloroplasts, mitochondria, nuclei, peroxisomes, cytoplasm, and apoplasts (Alscher et al., 2002; Gill and Tuteja, 2010). In plants, the role of SOD during environmental adversity, has received much attention since reactive oxygen species have been found to be produced during many stress conditions.

Here, we review and discuss three main points: The first one is about SODs, their types and significance; the second one illustrates SODs modulation in abiotic-stressed plants; and finally the third one reporting availability of SODs on genetic engineering in plants.

2. Superoxide Dismutase (SOD)

Superoxide dismutase (SOD, EC 1.15.1.1) is one of the most effective components of the antioxidant defense system in plant cells against ROS toxicity. SOD was first isolated from bovine blood as a green Cu-protein and was believed to function in Cu-storage (Mann and Keilin, 1938). Until reported in plants (McCord and Fridovich, 1969), SOD was recognized as a group of metallo-proteins having no know function. Available in most cells, and categorized into three main groups on the basis of the metal cofactor at active site, SODs are believed to occur in all oxygen metabolizing cells and also in all sub-cellular compartments (such as chloroplasts, mitochondria, nuclei, peroxisomes, cytoplasm, and apoplasts) (Alscher et al., 2002; Fink and Scandalios, 2002). SODs are metalloproteins working with Cu, Zn, Mn or Fe as cofactors (Mittler, 2002). Accordingly, three isoenzymes can be separated in plants having different localizations, structures and functions: Cu/Zn-SOD, Mn-SOD and Fe-SOD (Figure 1, Table 1; Alscher et al., 2002; Gill and Tuteja, 2010; Mahanty et al., 2012). Mn-SOD is present in the mitochondria and peroxisomes, while Cu/Zn-SOD is mainly cytosolic, mitochondrial and plastidic. Similarly, Fe-SOD seems to be frequent as its localization has been described in the chloroplasts, cytosol, mitochondria and peroxisomes (Giannopolitis and Ries, 1977; Inzé and Van Montagu, 1995; Arora et al.,

2002). Mn- and Fe-SODs are thought to be the most ancient types for their occurrence in both prokaryotes and eukaryotes. However, Cu/Zn-SOD has been detected predominantly in eukaryotes. The sensitivity of the isoenzymes is different too. Fe-SOD is resistant to KCN but can be inhibited by H_2O_2. Structurally it has homodimer and tetramer forms. Mn-SOD can be homodimeric and homotetrameric and is resistant to both KCN and H_2O_2, while Cu/Zn-SOD can be activated by both inhibitors (Alscher et al., 2002; Gill and Tuteja, 2010; Sharma et al., 2012). It has been reported that, like others antioxidant enzymes such as GR or APX, the activity of SOD also can be regulated by abscisic acid via gene expression (Gepstein and Glick, 2013). Information specific to major SOD types are briefly discussed in the following subsections.

Table 1. Types of SOD, Subcellular Location and Sensitivity

SOD Isoenzymes	Species	Structure	Subcellular Location	Sensitivity
Cu/Zn-SOD	*Pisum sativum* *Zea mays* *Spinacia oleracea* *Avena sativa* *Citrillus lanatus*	Homodimeric and Homotetrameric	Cytosol Chloroplast Peroxisome Mitochondria	H_2O_2 and KCN
Mn-SOD	*Vigna mungo* *Zea mays* *Pisum sativum* *Spinacia oleracea* *Nicotiana plumbaginifolia*	Homodimeric and Homotetrameric	Mitochondria Peroxisome	Not to H_2O_2 and KCN
Fe-SOD	*Arabidopsis thaliana* *Glycine max* *Oryza sativa* *Ginko biloba* *Zea mays* *Brassica campestris*	Homodimeric and tetrameric	Cytosol Chloroplast Peroxisome Mitochondria	H_2O_2 but not to KCN

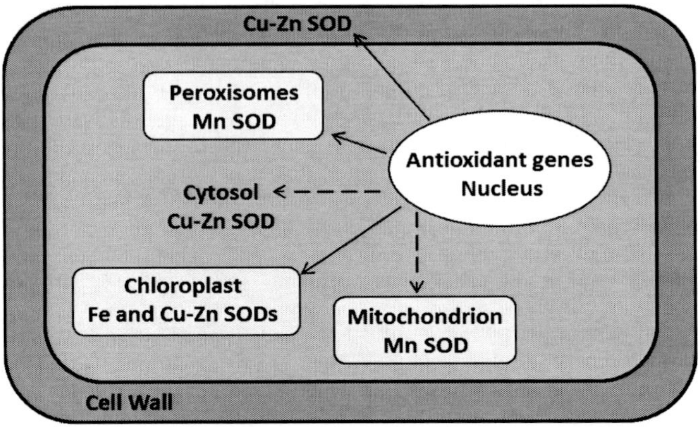

Figure 1. Localization of superoxide dismutases in plant cell.

2.1. Copper-Zinc Superoxide Dismutase

Cu/Zn-SODs, the most abundant class of plant SODs, can be homodimeric (periplasmic and cytosolic; with a molecular weight around 32 kD) and/or homotetrameric (chloroplastic and extracellular) (Alscher et al., 2002). However, plants can also exhibit monomeric forms of Cu/Zn-SODs (Kanematsu and Asada 1989; Schinkel et al., 2001). All cellular compartments can exhibit Cu/Zn-SODs; however, they mainly occur in cytosol, chloroplasts, peroxisomes, apoplasts, and nucleus (Karpinska et al., 2004; Karlsson et al., 2005; Kasai et al., 2006; Kim et al., 2008). The sensitivity of cytosolic Cu/Zn-SOD to H_2O_2 can be higher than chloroplastic Cu/Zn-SOD (Baum et al., 1983; Kwiatowski and Kaniuga, 1986). In Cu/Zn-SODs, bivalent Cu and Zn atoms are connected by a disulfide bond maintaining a non-covalent association (Fridovich, 1989). However, out of these two atoms, Cu is more important than Zn and hence, its substitution with another metal, or exclusion from the structure, causes enzyme inactivation (Bordo et al., 1994; Marino et al., 1995). On the other hand, the role of Zn is structural rather than functional because many reports indicated that substitution of Zn^{2+} with Co^{2+}, Hg^{2+}, Cd^{2+}, or Cu^{2+}, or even complete elimination of Zn could not significantly affect the activity of SOD (Cudd and Fridovich, 1982; Bordo et al., 1994; Marino et al.,

1995). However, Zn^{2+} helps in maintaining the structural stability of the active site of Cu/Zn-SOD (Marino et al., 1995). Recently, a profound effect of the presence of Zn in medium was reported on the synthesis of chloroplastic recombinant Cu/Zn-SOD (Tuteja et al., 2015).

2.2. Iron and Manganese Superoxide Dismutases

Fe-SOD is the first SOD where the Fe was used at its active site as the first metal co-factor. In evolutionary point of view, Fe-SOD is considered as a major constituent of the most ancient SOD group found both in prokaryotes and eukaryotes, and is thought to originate in the plastid before moving to the nuclear genome (Alscher et al., 2002; Fink and Scandalios, 2002). Studies on plants have evidenced chloroplast as the location of Fe-SOD. However, in addition to a typical chloroplastic Fe-SOD gene, cytosolic localization of Fe-SOD gene has been recently evidenced in *Pogonatum inflexum* (Kanematsu et al., 2013). This enzyme is resistant to KCN but is inactivated by H_2O_2 (Alscher et al., 2002). Fe-SOD has been reported to protect chloroplast nucleotides against oxidative stress, and was also considered essential for chloroplast development in *Arabidopsis* (Myouga et al., 2008). Coming to highlight major features of Mn-SODs, these enzymes carry only one metal atom per subunit, can be both homodimeric and homotetrameric with one Mn (III) atom per subunit, occur in mitochondria and peroxisomes, are resistant to KCN or H_2O_2, and cannot function without the Mn atom present at their active site (reviewed by Alscher et al., 2002). Notably, the restoration of Mn-SOD is not possible with Fe^{2+} despite the high similarity of Mn-SOD with Fe-SOD in their primary, secondary, and tertiary structures (Fridovich, 1986).

3. Response of SOD to Abiotic Stress

Abiotic stresses such as heat, cold, drought, salinity, and chemical contaminants represent some of the main factors or elements limiting agricultural productivity. Thus, the understanding of the plant responses to these abiotic stresses have become a pre-requisite in order to develop crop plants tolerating abiotic stresses (Tuteja and Gill, 2013). Nevertheless, as an important component of plant defense machinery within a cell, SODs constitute the first defense line against abiotic stress-accrued enhanced ROS and its reaction products where SODs catalyze the dismutation of $O_2^{\bullet-}$ to H_2O_2 and O_2. Here-under, recent reports available on the modulation of SODs in abiotic-stressed plant species are critically discussed.

3.1. Salt Stress

Salinity is considered as one of the most serious abiotic stress factors, sometimes combined with drought in arid and semi-arid fields, which may influence the crops quality and amount (Mittler, 2006). Several plants (halophytes) have the ability to tolerate high soil salt concentration or water deficit but most of the plants (including crops) are non-tolerant and show decreased photosynthetic activity or morphological disorders (Dat et al., 2000; Meloni et al., 2003; Sheokand et al., 2008; Szepesi et al., 2008; Zsigmond et al., 2012; Duque et al., 2013). All the studies confirm that salinity, like other abiotic stressors, evokes the overproduction of ROS and changes the activities of the ROS-scavenging enzymes. Some of the results confirmed that salt stress triggers an increase in the total SOD activity while other data showed no change or the opposite tendency with respect to either total SOD activity or that of the isozymes (Table 2). For instance, SOD activity showed an increase in the leaves of *Cicer arietinum*, *Beta vulgaris*, and *Brassica juncea* but was diminished in those of *Vigna unguiculata* (Hernandez et al., 2006; Bor et al., 2003; Sheokand et al., 2008; Kumar et al., 2013; Rasool et al., 2013). Moreover, even in the seedlings of *Arabidopsis* and *Nicotiana* elevated SOD activity was detected (Zsigmond et al., 2012; Lee et al., 2013).

Table 2. Changes of SOD Activity under Salt Stress, Drought, Chilling and Heat Stress

Type of Abiotic Stress Factor	Plant Name	Type of SOD Activated	Plant Organe investigated	Change in SOD Activity*	References
Salinity (NaCl)	*Arabidopsis thaliana* L.	Total SOD	Seedlings	++	Zsigmond et al., 2012
	Cicer arietinum L. cv. SKUA- 06 and -07	Total SOD	Leaves	++	Rasool et al., 2013
	Beta vulgaris L. cv. Ansa	Total SOD	Leaves	++	Bor et al., 2003
	Vigna unguiculata (L.) Walp.	Total SOD	Leaves	-	Hernandez et al., 2006
	Nicotiana tabacum cv. Xanthi	Cu/Zn-SOD	Seedlings	++	Lee et al., 2013
Drought (water Deficit)	*Oryza sativa* L. cv. Malviya-36 and Pant-12	Total SOD	Roots	++	Sharma and Dubey, 2005
	Olea europaea L. cv. Koroneiki	Total SOD	Leaves	++	Doupis et al., 2013
	Gossypium herbaceum genotype RAHS 187	Total SOD	Leaves	++ Later - -	Deeba et al., 2012
	Triticum aestivum L. cv. Centauro	Total SOD	Leaves	- -	Alexieva et al., 2001
Low Temperature (Chilling)	*Nicotiana tabacum* cv. HHDJY	Total SOD	Shoot	No Change	Gechev et al., 2003
	Oryza sativa L. cv. Xiangnuo-1 (tolerant)	Total SOD	Shoot	No Change	Huang and Guo, 2005
	Hordeum vulgare L. cv. Akhisar	Total SOD	Leaves	- -	Mutlu et al., 2013
	Pennisetum glaucum L.	Total SOD	Leaves	- -	Mahanty et al., 2012
High Temperature (Heat)	*Festuca arundinacea* L.	Total SOD	Leaves	No Change	Jiang and Huang, 2001
	Jatropha curcas L.	Total SOD	Leaves	++	Silva et al., 2013
	Poa pratensis L.	Total SOD	Leaves	No Change	Jiang and Huang, 2001

* (++) indicates significant and (+) indicates non significant increase, while (- -) refers to significant decrease and (-) signals non significant reduction in enzyme activity.

Considerable variations in SOD activity in response to salt stress are evident at inter-specific or intra-specific level. The activity of SOD and its isoform Cu/Zn-SOD increased in both the salt tolerant and sensitive cultivars of *Oryza sativa* against salinity (Mishra et al., 2013). Earlier also, it has been shown that salinity increases SOD activity in salt-tolerant cultivars and decreases it in salt-sensitive cultivars, in both leaves (Dionisio-Sese and Tobita, 1998; Hernandez et al., 2000) and roots (Shalata et al., 2001). While examining the long-term effects of salt stress in two salt-tolerant lines (Kharchia65, KRL19) and two salt-sensitive lines (HD2009, HD2687) of *Triticum aestivum*, Sairam et al. (2005) observed the exhibition of a higher increase in SOD activity by the salt tolerant line Kharchia65 when compared to the salt sensitive line HD2687. These contradictions suggest that SOD activity is affected by several conditions, depending on the plant species (tolerant or not), the duration and the concentration applied and, furthermore, the plant organ used for the measurements. The authors detected high Mn-SOD activity also in cytosolic fraction. However, Mn-SOD activity in both fractions was higher in Kharchia65, and further increased under salt stress. Cytosolic Mn-SOD activity was very rudimentary, suggesting that cytosolic Mn-SOD has very little role in the scavenging of salinity induced production of O_2^- radical. Additionally, the authors observed Cu/Zn-SOD in all the three fractions, the highest being in chloroplast followed by mitochondria and cytosol. In Kharchia65, though salinity increased the Cu/Zn-SOD activity in both cytosolic and chloroplastic fractions, in mitochondria, there was very little or no increase in Cu/Zn-SOD activity due to salinity but this genotype maintained higher activity than HD2687. Concerning Fe-SOD, it was predominantly present in chloroplastic fraction, while some activity was also detected in cytosolic and mitochondrial fractions. Here also, a higher activity was observed in Kharchia65, both in control and salt-stressed plants than HD2687. Under salt stress, Fe-SOD activity increased only slightly in cytosolic and mitochondrial fractions and significantly in chloroplastic fraction. Salinity-induced increase in chloroplastic Fe-SOD has been reported in *Pisum sativum* (Gomez et al., 1999).

3.2. Drought Stress

In addition to salinity, water deficit, may have an effect on the productivity of crops. Actually, it is not a marginal question, as the weather has become more changeable and unpredictable. Thus, acclimation to drought is at a great challenge for non-tolerant plants. In this case, we note that it is well known that drought stress generates numerous physiological alterations in plants such as a decline in CO_2 fixation due to abscisic acid-mediated stomatal closure, which leads to decreased the photosynthetic rate, and/or morphological disorders (mainly reduced organs growth) can be observed (Alexieva et al., 2001). Moreover, elevated ROS production arises, accompanied by lipid peroxidation and changes in both enzymatic and non-enzymatic antioxidants (Dat et al., 2000; Arora et al., 2002; Csiszar et al., 2007; Tari et al., 2008; Gill and Tuteja, 2010; Dinakar et al., 2012; Sen and Alikamanoglu, 2013). In plants exposed to water stress, the total SOD activity seemed usually to increase in the leaves of *Olea europaea* or in the shoots of *Oryza sativa* cultivars in both tolerant and non-tolerant tomato species (Table 2; Sharma and Dubey, 2005;Unyayar et al., 2005; Doupis et al., 2013). But in the leaves of *Gossypium* or two grasses (*Festuca arundinacea* and *Poa pratensis*) elevated activity of total SOD was followed by diminution (Jiang and Huang, 2001; Deeba et al., 2012). At the same time, there are available data on a remarkable decrease in SOD activity in pea and wheat (Alexieva et al., 2001). In *Triticum aestivum*, SOD activity increased or remained unchanged in the early phase of drought but decreased with prolonged water stress (Zhang and Kirkham, 1995). SOD can protect photosystem II against $O_2^{\bullet-}$ generated by oxidative and water stress (Martinez et al., 2001). In a recent study, the Tibetan wild *Hordeum vulgare* genotypes XZ16 and XZ5 were reported to exhibit significantly increased SOD activity under drought treatments during anthesis period (Ahmed et al., 2013).

3.3. Cold Stress

Cold can provoke functional disorders in the chloroplasts or mitochondria in sensitive plants, which unequivocally results in ROS-overproduction, i.e., oxidative stress (Prasad et al., 1994; Rivero et al., 2001). Earlier experiments indicated that remarkable H_2O_2 formation was found to occur in stressed plants during chilling and higher antioxidant enzyme activities seemed to correlate with higher tolerance (Dat et al., 2000; Lee and Lee, 2000; Huang and Guo, 2005).

In tolerant tobacco and rice cultivars, there were no changes in the total SOD activity, while in the shoots or leaves of the sensitive types of banana, tobacco and rice. It significantly decreased, similar to that of both sensitive and tolerant cultivars of barley (Table 2; Kang et al., 2003; Huang and Guo, 2005; Xu et al., 2010; Mutlu et al., 2013), which also supports the positive correlation between higher antioxidant enzyme activity (including SOD) and chilling tolerance.

3.4. Heat Stress

Extremely high temperatures may cause significant physiological and morphological destructions in plants combined with water deficit or salt stress (Mittler, 2006). Though there is a complex system for thermo-tolerance in plants comprising heat shock proteins regulated by heat stress transcription factors, elevated temperatures usually trigger oxidative stress as well (Dat et al., 2000; Kotak et al., 2007; Ceylan et al., 2013). It has been reported that accumulation of H_2O_2 occurs during heat stress and total phenol content increases, while the activity of peroxidase diminishes in tomato exposed to higher temperatures (Dat et al., 2000; Rivero et al., 2001). We have relatively few data on the changes of SOD activity in heat-stressed plants. In two grasses (*Festuca arundinacea* and *Poa pratensis*) total SOD activity showed no alterations, while in the leaves of mulberry and *Jatropha curcas* it increased significantly, which resulted in the accumulation of H_2O_2 as it was published earlier (Table 2; Chaitanya et al., 2002; Jiang and Huang, 2001; Silva et al., 2013).

3.5. Heavy Metals

In recent years the amount of heavy metals has been elevated due to mining, industrial, and agricultural activities and the output of municipal wastes, which may result in the overproduction of ROS. Heavy metal stress is generally followed by physiological and/or morphological alterations in plants. Typical symptoms are chlorosis, reduced germination rate, retarded growth of vegetative organs and at toxic level necrosis of plant tissues (Dat et al., 2000; Arora et al., 2002). At the cellular level, not only toxic metals such as Pb, Hg, and Cd can cause oxidative stress in the plant cells but also essential Fe, Cu or Zn at higher concentrations than physiological level might trigger the increase of ROS generation. The response of SOD to metal/metalloid stress varies considerably depending upon the plant species, developmental stage, metal in the experiment and the exposure time.

Copper (Cu) is an important biometal and its higher accumulation can cause lipid peroxidation; because of his redox activity he catalyze $^{\bullet}OH$ radical formation through the Fenton reaction and alter the activity of antioxidant enzymes (Arora et al., 2002; reviewed by Schutzendubel and Polle, 2002; Szollosi et al., 2011). Literature data show that the activity of either total SOD or Cu/Zn-SOD in leaves or roots is generally induced by excess copper (Table 3). Increasing Cu concentrations up to 1.5 mM caused increased SOD activity in the leaves of *Zea mays* cultivars (3223 and 31G98) (Tanyolac et al., 2007).

Table 3. Changes of SOD Activity under Heavy Metal Stress

Type of Abiotic Stress Factor (Heavy metals)	Plant Name	Type of SOD Activated	Plant Organe investigated	Change in SOD Activity*	References
Cu	*Arabidopsis thaliana* L.	Total SOD	Leaves	++	Drazkiewicz et al., 2004
	Brassica juncea L.	Total SOD	Seedlings	++	Szoll osi et al., 2012
	Pisum sativum L.	Total SOD	Leaves	++	Chaoui and El Ferjani, 2005
	Elsholtzia haichowensis	Cu/Zn-SOD	Roots	++	Zhang et al., 2008
Zn	*Brassica napus* L. cv. Suyou N°.1	Total SOD	Leaves	-	Wang et al., 2009
	Brassica juncea L.	Total SOD	Shoots	++	Prasad et al., 1999
	Eruca sativa L. (Rocket)	Total SOD	Leaves	- -	Ozdener and Aydin, 2010
Fe	*Helianthus annuus* L. cv. Mycosol 2	Total SOD	Leaf disc	- -	Gallego et al., 1996
	Nicotiana plumbaginifolia L.	Total SOD	Leaves	No Change	Kampfenkel et al., 1995
	Phaseolus vulgaris L. cv. Pinto	Fe-SOD	Leaves	- -	Shainberg et al., 2000
		Mn-SOD	Leaves	No Change	
		Cu/Zn-SOD	Leaves	No Change	
Mn	*Vitis vinifera* L.	Total SOD	Leaves	-	Mou et al., 2011
	Oryza sativa L. cv. Pant-12	Mn-SOD	Shoot	++	Srivastava and Dubey, 2011
		Cu/Zn-SOD	Shoot	++	
		Fe-SOD	Shoot	++	
Cd	*Arabidopsis thaliana* L.	Total SOD	Leaves	-	Cho and Seo, 2005
	Vigna radiata L.	Total SOD	Leaves	++	Anjum et al., 2011
	Brassica juncea L.	Total SOD	Leaves	+	Kumar et al., 2013
	Triticum durum Desf cv. Adamello	Total SOD	Leaves	- -	Milone et al., 2003
Ni	*Alyssum argenteum* L.	Total SOD	Shoots	-	Schickler and Caspi, 2002
	Zea mays L. cv. 777	Total SOD	Leaves	++	Kumar et al., 2007
	Triticum aestivum L. cv. Zyta	Total SOD	Shoot	- -	Gajewska and Skłodowska, 2008

Zinc (Zn) is also an essential microelement for plants; Zn is involved in N-metabolism, cell growth regulation, auxin biosynthesis and acts as a cofactor of several enzymes such as oxido-reductases, tranferases, and ligases (Rout and Das, 2003; Broadley et al., 2007). At the same time, it may cause oxidative stress, and consequently cell wall and membrane damage, lipid peroxidation or disorders in various metabolic processes (Broadley et al., 2007; Janas et al., 2010; Yadav, 2010). Plants exposed to relatively high zinc contents (1.12 mM) show chlorosis and growth inhibition (Wang et al., 2009). Similar to Cu, zinc is also a bivalent cation but redox inactive; thus, it evokes ROS overproduction indirectly, not via the Fenton reaction causing lipid peroxidation as well (Dat et al., 2000; Cuypers et al., 2002). According to data published in the literature, SOD activity seems to be dose and time-dependent (Table 3) (Prasad et al., 1999; Rao and Sresty, 2000; Wang et al., 2009; Ozdener and Aydin, 2010).

Iron (Fe) is an essential heavy metal for plants, involved in photosynthesis, N-fixation, hormone biosynthesis, and respiration and is the cofactor of antioxidant enzymes such as Fe-SOD, catalase, ascorbate peroxidase or guaiacol peroxidase (Becana et al., 1998; Shainberg et al., 2000; Arora et al., 2002). Excess Fe may result in bronzing and brownish pigmentation of leaves as visual symptoms. As a transitional metal, it induces ROS generation through the Fenton reaction and lipid peroxidation (Dat et al., 2000; Fang et al., 2001). Results which have been reported up to now show that total SOD activity may be reduced or constant in leaves but Fe-SOD can become induced after excess Fe, such as in those of bean (Table 3; Kampfenkel et al., 1995; Gallego et al., 1996; Shainberg et al., 2000; Sinha and Saxena, 2006).

There are relatively few data available on manganese (Mn) toxicity in plants, although this transitional metal is the cofactor of Mn-SOD and plays a key role in respiration, tricarboxylic acid cycle and N-metabolism (Alscher et al., 2002; Bueno and Piqueras, 2002; Xue et al., 2004). Relatively high (3 and 6 mM) Mn concentrations induced the activity of total SOD and SOD isoforms in the shoots of rice, while 60 mM Mn partially inhibited SOD (Table 3; Mou et al., 2011; Srivastava and Dubey, 2011).

Cadmium (Cd), known as a non-essential toxic heavy metal, generates various morphological and physiological changes and disorders in plants. Reduction of plant growth and photosynthesis, chlorosis in leaves or inhibition of seeds germination are typical signs of Cd toxicity (Chaoui et al., 1997; Arora et al., 2002; Benavides et al., 2005; Yadav, 2010; Kranner and Colville, 2011). These symptoms are the probable consequences of oxidative stress, i.e., formation of ROS, increased lipid peroxidation (LP), and decreased or increased activity of enzymatic and non-enzymatic antioxidants (Gallego et al., 1999; Hasan et al., 2009; Szollosi et al., 2009). Cadmium (Cd) mediated enhanced SOD activity has been reported in a number of plants including *Bacopa monnieri* (Mishra et al., 2006), *Brassica juncea* (Mobin and Khan, 2007), *Vigna mungo* (Singh et al., 2008), *Vigna radiate* (Anjum et al., 2008a), *Triticum aestivum* (Khan et al., 2007), *Brassica campestris* (Anjum et al. 2008b), *Zea mays* (Ekmekci et al., 2008), *Pisum sativum* (Agrawal and Mishra, 2009) and *Oryza sativa* (Wu et al., 2006). In contrast, Cd-accrued significantly decreased SOD activity exhibition were reported in *Arabidopsis thaliana* (Cho and Seo, 2005), *Triticum durum* (Milone et al., 2003), *Glycine max* (Noriega et al., 2007), and *Capsicum annuum* (Table 3) (Leon et al., 2002).

Nickel (Ni) is also a transitional metal and an essential microelement for plants (Gajewska et al., 2006). At the same time, high levels of Ni might evoke growth inhibition, chlorosis, necrosis and lipid peroxidation or increased ROS production in the plant cells (Gajewska et al., 2006; Kumar et al., 2007). Micromolar concentrations of Ni seemed to induce SOD in maize shoots and leaves, while millimolar levels triggered a decrease in vegetative organs of *Alyssum argenteum* and pea (Table 3; Schickler and Caspi, 2002; Gajewska and Skłodowska, 2005; Kumar et al., 2007). Treatment of *Triticum aestivum* plants with high Ni concentration resulted in a decrease in SOD activity (Gajewska and Skłodowska, 2008) and in hairy root of *Alyssum bertolonii* and *Nicotiana tobaccum* (Boominathan and Doran, 2002) but an opposite effect has been observed in *Zea mays* leaves under Ni exposure (Kumar et al., 2007).

3.6. UV B-Radiation

The stratospheric ozone depletion encoded an increase in UV-B radiation (280-320) and affected harmfully the living organisms (Mackerness, 2000). It is also well-known that the UV-B influences plant growth and some other metabolic processes. Indeed, it may alter the pigment composition and therefore the efficiency of photosynthesis one (Balakumar et al., 1997; Mackerness, 2000; Costa et al., 2002). Moreover, it can cause damage to proteins and DNA since it induces ROS formation and affects the activity of antioxidant enzymes such as GR or GPX (Dat et al., 2000). In accordance with this, the UV-B irradiated plants have in general significantly elevated SOD activity (total and isoforms) in the leaves, except for *Nicotiana plumbaginifolia* (Willekens et al., 1994; Rao et al., 1996; Dai et al., 1997; Alexieva et al., 2001).

3.7. Air Pollutants

At the end of the 20th century it was proven that air pollutants may be very harmful not only to humans but also to plants. This problem has not been solved Up to now, and results from industrial activities, especially those using machines working with fossil fuel, etc. Indeed, the emission of pollutant airs has not been reduced. In addition, the ozone (O_3) and the sulphur dioxide (SO_2) are considered to be the main atmospheric pollutants. The both can enter to plants through the stomata and can easily cause physiological and morphological injuries in the tissues then in the crop yield (Dat et al., 2000; Alonso et al., 2001). Ozone toxicity is related to the formation and accumulation of ROS such as $^{\bullet}O_2$ radicals and H_2O_2 (Rao et al., 1996). Enzymatic antioxidants are involved in the avoidance of oxidative burst caused by ozone. SOD is increased in the leaves of *Arabidopsis*, rice and the younger needles of *Pinus halepensis*, which suggests that the defense system is probably affected by the age of the plant (Rao et al., 1996; Alonso et al., 2001; Ismail and Suroto, 2012). In *Zea mays* seedling, acute (single) or chronic (3, 6, and 10 consecutive days) exposure led to exhibition of increases in transcript levels of mitochondrial Mn-SOD and cytosolic Cu/Zn-SODs in leaves (Ruzsa et al., 1999). However, transcript level of chloroplastic Cu/Zn-SOD was evidenced down-regulated.

SO_2 entering the leaf becomes hydrated and forms sulphur trioxide (HSO_3-), $^{\bullet}O_2$- and $^{\bullet}$OH radicals (Dat et al., 2000; Bernardi et al., 2001). Thus, sulphur dioxide can be regarded as a hazardous pollutant. In order to scavenge these reactive radicals, antioxidant molecules are involved. For instance, the induction of total SOD and Fe-SOD activities has been recorded in the leaves of bean and tolerant pea (Madamanchi and Alscher, 1991; Bernardi et al., 2001).

4. GENETIC ENGINEERING OF PLANT SOD

The induction of SOD in response to diverse environmental conditions discussed in the previous section indicates that it plays an important role in plant's defense mechanism. It may be a crucial component and a veritable platform, in which case its genetic manipulation might result in stress-tolerant phenotypes. Alternatively, collected data may simply demonstrate the ubiquity of oxidative stress in plant processes. In this case, alteration of SOD activity resulting from their overproduction or underproduction would perturb the normally optimized mechanisms. Using transgenic approaches, several species were studied aiming at the improvement of tolerance to stress enhancing the antioxidant gene capacities. Indeed, Table 4 shows some examples of the positive response obtained with regard to increased tolerance to various abiotic stresses like cold, drought, heat, salt, hydrogen peroxide, methyl viologen, and metals stresses. Wang et al. (2005) evidenced the improved ability of *Oryza sativa* against methyl viologen (MV) and drought stress where the *Pisum sativum* Mn-SOD gene was expressed in *Oryza sativa* chloroplast. In another instance, the overexpression of Cu/Zn-SOD gene in *Nicotania tabacum* plants was confirmed to improve its resistance to oxidative stress (Gupta et al., 1993), and tolerance to MV and pure cercosporin was achieved in sugar beet by transferring a *Lycopersicon esculentum* SOD gene (Tertivanidis et al., 2004). Introduction of SOD genes into plants has been evidenced to help plants to more efficiently eliminate ROS under MV or environmental stresses such as chilling, ozone, water deficit, and salt stresses (Foyer et al., 1994; Bowler et al., 1991; McKersie et al., 1993, 1999; Gupta et al., 1993; Van Camp et al., 1996; Wang et al., 2005). However, no improvements in stress tolerance were observed in several transgenic plants with extra-genetic SODs (Tepperman and Dunsmuir, 1990; Sunkar et al., 2006; Jia et al., 2009). The complexity of the ROS detoxification system as well as the differences of SOD isoenzymes have been advocated as the major factors in this context (Kwon et al., 2002; Mylona and Polidoros, 2010). Wang et al. (2004) constructed transgenic *Arabidopsis* overexpressing Mn-SOD where, the activity of Mn-SOD as well as that of Cu/Zn-SOD and Fe-SODs was higher than that of wild type. Additionally, when treated with 150 mM NaCl, the transgenic plants grew well and exhibited lower lipid peroxidation product (MDA), while the wild type plants withered gradually, this indicated the enhanced salt tolerance of transgenic *Arabidopsis*. The studies on transgenic plants with suppressed SOD have provided substantial evidences of $O_2^{\bullet-}$ regulation of antioxidant gene expression. In this context, Rizhsky et al. (2003) performed the transcriptome analysis of knockdown *Arabidopsis* plants with suppressed expression of chloroplastic Cu/Zn-SOD (CSD2) and reported the exhibition of induced chloroplast and nuclear encoded genes as a result of $O_2^{\bullet-}$ accumulation under optimal conditions (Rizhsky et al., 2003). Recently, in order to investigate new gene resource for enhancing *Oryza sativa*-tolerance to salt stress, Mn-SOD gene from halophilic archaeon (*Natrinema altunense* sp.) (NaMn-SOD) was isolated and introduced into *Oryza sativa* cv. Nipponbare by *Agrobacterium*-mediated transformation (Chen et al., 2013). The transformants (L1 and L2) showed some NaMn-SOD expression and increased total SOD activity, which contributed to higher efficiency of ROS elimination under salt stress. The levels of $O_2^{\bullet-}$ and H_2O_2 were significantly decreased. In addition, they exhibited higher levels of photosynthesis, whereas lower relative ion leakage and MDA content compared to wild-type plants. The authors were able to obtain transgenic seedlings that were phenotypically healthier, where the heterologous expression of NaMn-SOD improved salt tolerance in *Oryza sativa* was evidenced by the exhibition of significantly decreased levels of $O_2^{\bullet -}$ and H_2O_2. In contrast to the discussed above reports, where the overexpression of SOD in plants was reported to render them stress-tolerant, in some transgenics, the overexpression of cytosolic or chloroplastic SOD provided only moderate or minimize tolerance, attributed to the type of overexpressed SOD and its subcellular localization (Allen et al., 1997). In particular, the inactivation of Cu/Zn-SOD and Fe-SOD by H_2O_2, their reaction end product, and the contesting resistance of Mn-SOD to H_2O_2 may be pertinent to their particular effects. In this case, some approaches like antisense technology is feasible now and should be complementary to any overexpression studies.

Table 4. Some examples of the transgenic plants with potential stress tolerance expressing SOD

Gene	Native species	Target species	Stress tolerance	Reference
Superoxide dismutase	*Oryza sativa*	*Nicotiana tabacum*	salinity, water, PEG treatment	Badawi et al. 2004
	Tamarix androssowii	*Populus davidiana x P. bolleana*	salinity	Wang et al. 2010
	Pisum sativum	*Oryza sativa*	drought	Wang et al. 2005
	Arachis hypogaea	*Nicotiana tabacum*	salinity, drought	Neji et al. 2015
Superoxide dismutase/Catalase	*Gossypium hirsutum*	*Gossypium hirsutum*	salinity, methyl viologen	Luo et al. 2013
Superoxide dismutase/Ascorbate peroxidase	*Rheum austral/Potentilla astrisanguinea*	*Arabidopsis thaliana*	Cold, methyl viologen	Shafi et al. 2014
	Manihot esculenta	*Manihot esculenta*	hydrogen peroxide, cold	Xu et al. 2014
	Solanum tuberosum	*Solanum tuberosum*	heat, methyl viologen	Tang et al. 2006

CONCLUSION

Plants are sessile and always exposed to abiotic stress factors, so it is a challenge for them to be tolerant to the unfavorable conditions, since these factors often overlap and result in more serious crops damages (Mittler, 2006). Various studies and investigations have already proved that the ROS generation and the redox deviations caused by different abiotic stress factors are the consequence of enzymatic disorders, reduced photosynthesis, and damage of proteins, lipids and DNA, which affect the plant growth and also the low crop yield. The complexity of the antioxidant defense system requires basic research to continue, but it is also important to summarize the results from time to time. Hence, change in SOD activity under abiotic stress conditions seems to be contradictory. This phenomenon argues for multifactorial dependence of the enzyme activity such as plant species, plant organs investigated, concentration applied or time of exposure. Although there are always popular topics in plant sciences in which stress physiology is one of them, it would still be useful to refresh the data on, for example, oxidative stress triggered by air pollutants or UV-B radiation. Naturally, it is indispensable to work on transgenic varieties to understand and to improve stress tolerance (Potters et al., 2007; Gill and Tuteja, 2010). Little is known about the regulation of SOD types by post-translational modifications (Holzmeister et al., 2015). If performed, these studies can help us in producing transgenic plants exhibiting suppressed SODs, and high capacity for the regulation of elevated cellular $O_2^{\bullet-}$. Metallic nanoparticles are being rapidly contaminating varied environmental compartments, and these nanoparticles can be accumulated in plants. Metallic nanoparticles such as nanoscale Cu, Fe and Zn may interact with SODs and modulate their biochemistry and functions since Cu, Fe and Zn are structural constituents of SODs. Hence, studies on the potential interaction of SOD isoforms (Cu/Zn-SOD, Mn-SOD, Fe-SOD) with Cu, Fe, and Zn nanoparticles can be promising under the umbrella of "nanometallomics," Additionally, more exhaustive research on the "SOD isoform-bioinformatics" can confirm potential antiquity among SOD isoforms.

ACKNOWLEDGMENTS

This work was in part financially supported by a grant from the Tunisian Ministry of Higher Education and Scientific Research contract program_2015-2018 CBS-LBAP/code: LR15CBS03.

REFERENCES

Agrawal, S. B. and Mishra, S. (2009). Effects of supplemental ultraviolet-B and cadmium on growth, antioxidants and yield of *Pisum sativum* L. *Ecotoxicology Environmental Safety,* 72: 610–618.

Ahmed, I. M., Cao, F., Zhang, M., Chen, X., Zhang, G. et al. (2013). Difference in yield and physiological features in response to drought and salinity combined stress during anthesis in Tibetan wild and cultivated barleys. *PLoS ONE,* 8(10): e77869.

Alexieva, V., Sergiev, I., Mapelli, S. and Karanov, E. (2001). The effect of drought and ultraviolet radiation on growth and stress markers in pea and wheat. *Plant Cell Environment*, 24, 1337-1344.

Allen, R. D., Webb, R. P. and Schake, S. A. (1997). Use of transgenic plants to study antioxidant defenses. *Free Radical Biology and Medicine,* 23:73–479.

Alonso, R., Elvira, S., Castillo, F. J. and Gimeno, B. S. (2001). Interactive effects of ozone and drought stress on pigments and activities of antioxidative enzymes in *Pinus halepensis*. *Plant Cell Environment,* 24: 905-916.

Alscher, R. G., Erturk, N. and Heath, L. S. (2002). Role of superoxide dismutases (SODs) in controlling oxidative stress in plants. *Journal of Experimental Botany,* 53: 1331-1341.

Anjum, N. A, Umar, S., Ahmad, A., Iqbal, M. and Khan, N. A. (2008a). Ontogenic variation in response of *Brassica campestris* L. to cadmium toxicity. *Journal of Plant Interaction,* 3:189–198.

Anjum, N. A., Umar, S., Iqbal, M. and Khan, N. A (2008b). Growth characteristics and antioxidant metabolism of moongbean genotypes differing in photosynthetic capacity subjected to water deficit stress. *Journal of Plant Interaction,* 3:127–136.

Anjum, N. A., Umar, S., Iqbal, M. and Khan, N. A. (2011). Cadmium causes oxidative stress in mung bean by affecting the antioxidant enzyme system and ascorbate-glutathione cycle metabolism. *Russian Journal of Plant Physiology,* 58: 92-99.

Arora, A., Sairam, R. K. and Srivastava, G. C. (2002). Oxidative stress and antioxidative system in plants. *Current Science*, 82: 1227-1238.

Badawi, G. H., Yamauchi, Y., Shimada, E., Sasaki, R., Kawano, N., Tanaka, K. and Tanaka, K. (2004)). Enhanced tolerance to salt stress and water deficit by overexpressing superoxide dismutase in tobacco (*Nicotiana tabacum*) chloroplasts. *Plant Science,* 166(4): 919-928.

Balakumar, T., Gayathri, B. and Anbudurai, P. R. (1997). Oxidative stress injury in tomato plants induced by supplemental UV-B radiation. *Biologia Plantarum,* 39: 215-221.

Baum, J. A., Chandlee, J. M. and Scandalios, J. G. (1983). Purification and partial characterization of a genetically defined superoxide dismutase (SOD-1) associated with maize chloroplasts. *Plant Physiology*, 73:31–35.

Becana, M., Moran, J. F. and Iturbe-Ormaetxe, I. (1998). Iron-dependent oxygen free radical generation in plants subjected to environmental stress: toxicity and antioxidant protection. *Plant Soil,* 201:137-147.

Benavides, M. P., Gallego, S. M. and Tomaro, M. L. (2005). Cadmium toxicity in plants. *Brazilian Journal of Plant Physiology*, 17: 21-34.

Bernardi, R., Nali, C., Gargiulo, R., Pugliesi, C., Lorenzini, G. and Durante, M. (2001). Protein pattern and Fe-superoxide dismutase activity of bean plants under sulphur dioxide stress. *Journal of Phytopathology,* 149: 477-480.

Boominathan, R. and Doran P. M. (2002). Ni-induced oxidative stress in roots of the Ni hyper-accumulator, *Alyssum bertolonii*. *New Phytology,* 156: 205–215.

Bor, M., Ozdemir, F. and Turkan, I. (2003). The effect of salt stress in lipid peroxidation and antioxidants in leaves of sugar beet *Beta vulgaris* L. and wild beet *Beta maritima* L. *Plant Science*, 164: 77-84.

Bordo, D., Djinovic, K. and Bolognesi, M. (1994). Conserved patterns in the Cu-Zn superoxide dismutase family. *Journal of Molecular Biology,* 238:366–386.

Bowler, C., Slooten, L., Vandenbranden, S., De Rycke, R., Botterman, J. et al. (1991). Manganese superoxide dismutase can reduce cellular damage mediated by oxygen radicals in transgenic plants. *EMBO Journal,* 10: 1723–1732.

Broadley, M. R., White, P. J., Hammond, J. P., Zelko, I. and Lux, A. (2007). Zinc in plants. *New Phytology,* 173: 677-702.

Bueno, P. and Piqueras, A. (2002). Effect of transition metals on stress, lipid peroxidation and antioxidant enzyme activities in tobacco cell cultures. *Plant Growth Regulators,* 36: 161-167.

Ceylan, H. A., Turkan, I. and Sekmen, A. H. (2013). Effect of coronatine on antioxidant enzyme response of chickpea roots to combination of PEG-induced osmotic stress and heat stress. *Journal of Plant Growth Regulation,* 32: 72-82.

Chaitanya, K. V., Sundar, D., Masilamani, S., Ramachandra Reddy, A. (2002). Variation in heat stress-induced antioxidant enzyme activities among three mulberry cultivars. *Plant Growth Regulation,* 36: 75-180.

Chaoui, A. and El Ferjani, E. (2005). Effects of cadmium and copper on antioxidant capacities, lignifications and auxin degradation in leaves of pea (*Pisum sativum* L.) seedlings. *Comptes rendus Biologies,* 328: 23-31.

Chaoui, A., Mazhoudi, S., Ghorbal, M. H. and El Ferjani, E. (1997). Cadmium and zinc induction of lipid peroxidation and effects on antioxidant enzyme activities in bean (*Phaseolus vulgaris* L.). *Plant Science,* 127: 139-147.

Chen, Z., Pan, Y.H., An, L. Y., Yang, W. J., Xu, L. G. and Zhu, C. (2013). Heterologous expression of a halophilic archaeon manganese superoxide dismutase enhances salt tolerance in transgenic rice. *Russian Journal of Plant Physiology*, 60:359–366.

Cho, U. H., Seo, N. H. (2005). Oxidative stress in *Arabidopsis thaliana* exposed to cadmium is due to hydrogen peroxide accumulation. *Plant Science,* 168: 113-120.

Costa, H., Gallego, S. M. and Tomaro, M. L. (2002). Effect of UV-B radiation on antioxidant defense system in sunflower cotyledons. *Plant Science,* 162: 939-945.

Cuypers, A., Vangronsveld, J. and Clijsters, H. (2002). Peroxidases on roots and primary leaves of *Phaseolus vulgaris* copper and zinc phytotoxicity: a comparison. *Journal of Plant Physiology*, 159: 869-876.

Csiszar, J., Lantos, E., Tari, I., Madosa, E., Wodala, B., Vashegyi, A., et al. (2007). Antioxidant enzyme activities in Allium species and their cultivars under water stress. *Plant Soil Environment*, 53: 517-523.

Cudd, A. and Fridovich, I. (1982). Electrostatic interactions in the reaction mechanism of bovine erythrocyte superoxide dismutase. *Journal of Biological Chemistry,* 257:11443–11447.

Dai, Q., Yan, B., Huang, S., Liu, X., Peng, S., Miranda, M. L. L., et al. (1997). Response of oxidative stress defense systems in rice (*Oryza sativa*) leaves with supplemental UV-B radiation. *Physiolgy Plant,* 101: 301-308.

Dat, J., Vandenabeele, S., Vranova, E., Van Montagu, M., Inzé, D. and Van Breusegem, F. (2000). Dual action of the active oxygen species during plant stress responses. *Cell Molecular Life Science*, 57: 779-795.

Deeba, F., Pandey, A. K., Ranjan, S., Mishra, A., Singh, R., Sharma, Y. K., et al. (2012). Physiological and proteomic responses of cotton (*Gossypium herbaceum* L.) to drought stress. *Plant Physiology and Biochemistry,* 53: 6-18.

Dionisio-Sese, M. L. and Tobita, S. (1998). Antioxidant responses of rice seedlings to salinity stress. *Plant Science,* 135:1–9.

Dinakar, C., Djilianov, D. and Bartels, D. (2012). Photosynthesis in desiccation tolerant plants: energy metabolism and antioxidative stress defense. *Plant Science,* 182: 29-41.

Doupis, G., Bertaki, M., Psarras, G., Kasapakis, I. and Chartzoulakis, K. (2013). Water relations, physiological behavior and antioxidant defence mechanism of olive plants subjected to different irrigation regimes. *Science Horticulture,* 153: 150-156.

Drawzkiewicz, M., Skorzynska-Polit, E. and Krupa, Z. (2004). Copper-induced oxidative stress and antioxidant defence in *Arabidopsis thaliana. Bio Metals,* 17:379-387.

Duque, A. S., de Almeida, A. M., da Silva, A. B., da Silva, J. M., Farinha, A. P., Santos, D., et al. (2013). Abiotic stress responses in plants: unraveling the complexity of genes and networks to survive. In: Vahdati, K., Lesliepp, C. (Eds.), *Abiotic Stress: Plant Responses and Applications in Agriculture. In Tech*, pp. 49_101.

Ekmekci, Y., Tanyolac, D. and Ayhana, B. (2008). Effects of cadmium on antioxidant enzyme and photosynthetic activities in leaves of two maize cultivars. *Journal of Plant Physiology,* 165:600–61.1.

Fang, W. C., Wang, J. W., Lin, C. C. and Kao, C. H. (2001). Iron induction of lipid peroxidation and effects on antioxidative enzyme activities in rice leaves. *Plant Growth Regulation,* 35: 75-80.

Fink, R. C. and Scandalios, J. G. (2002). Molecular evolution and structure function relationships of the superoxide dismutase gene families in angiosperms and their relationship to other eukaryotic and prokaryotic superoxide dismutases. *Archives of Biochemystry and Biophysics,* 399:19–36.

Foyer, C. H., Descourvieres, P. and Kunert, K. J. (1994). Protection against oxygen radicals: an important defence mechanism studied in transgenic plants. *Plant Cell Environment,* 17:507–523.

Fridovich, I. (1986). Superoxide dismutases. *Advances in Enzymology and Molecular Biology,* 58:61–97.

Fridovich, I. (1989). Superoxide dismutase: an adaptation to a paramagnetic gas. *Journal of Biology and Chemistry*, 264:7761–7764.

Gajewska, E. and Skłodowska, M. (2005). Antioxidative responses and proline level in leaves and roots of pea plants subjected to nickel stress. *Acta Physiologica Plantarum,* 27: 329-340.

Gajewska, E. and Skłodowska, M. (2008). Differential biochemical responses of wheat shoots and roots to nickel stress: antioxidative reactions and proline accumulation. *Plant Growth Regulation,* 54: 179-188.

Gajewska, E., Skłodowska, M., Słaba, M. and Mazur, J. (2006). Effect of nickel on antioxidative enzyme activities, proline and chlorophyll contents in wheat shoots. *Biological Plant*, 50: 653-659.

Gallego, S. M., Benavıdes, M. P. and Tomaro, M. L. (1996). Effect of heavy metal ion excess on sunflower leaves: evidence for involvement of oxidative stress. *Plant Science,* 121: 151-159.

Gallego, S. M., Benavides, M. P. and Tomaro, M. L. (1999). Effect of cadmium ions on antioxidant defense system in sunflower cotyledons. *Biological Plant,* 42: 49-55.

Gechev, T., Willekens, H., Van Montagu, M., Inzé, D., Van Camp, W., Toneva, V., et al. (2003). Different responses of tobacco antioxidant enzymes to light and chilling stress. *Journal of Plant Physiology*, 160: 509-515.

Gepstein, S. H. and Glick, B. R., (2013). Strategies to ameliorate abiotic stress-induced plant senescence. *Plant Molecular Biology*, 82: 623-633.

Giannopolitis, C. N. and Ries, S. K. (1977). Superoxide dismutases I. Occurrence in higher plants. *Plant Physiology,* 59: 309-314.

Gill, S. S. and Tuteja, N. (2010). Reactive oxygen species and antioxidant machinery in abiotic stress tolerance in crop plants. *Plant Physiology and Biochemistry,* 48: 909-930.

Gomez, J. M., Hernandez, J. A., Jimenez, A., del Rio, L. A. and Sevilla, F. (1999). Differential response of antioxidative enzymes of chloroplasts and mitochondria to long-term NaCl stress of pea plant. *Free Radical Research,* 31(Suppl):11–18.

Gupta, A. S., Webb, R. P., Holaday A. S. and Allen R. D. (1993). Overexpression of superoxide dismutase protects plants from oxidative stress. *Plant Physiology,* 103:1067–1073.

Halliwell, B. and Gutteridge, J. M. C. (1989). *Free Radicals in Biology and Medicine*. Oxford, Clarendon Press.

Halliwell, B. (2006). Reactive species and antioxidants. Redox biology is a fundamental theme of aerobic life. *Plant Physiology*, 141:312-22.

Hasan, S. A., Fariduddin, Q., Ali, B., Hayat, S. and Ahmad, A. (2009). Cadmium: toxicity and tolerance in plants. *Journal of Environmental Biology,* 30: 164-174.

Hasanuzzaman, M., Hossain, M. A., da Silva J. A. T. and Fujita, M. (2012). Plant responses and tolerance to abiotic oxidative stress: antioxidant defense is a key factor. In: Bandi V, Shanker, A. K., Shanker, C., Mandapaka, M. (eds.) *Crop stress and its management: perspectives and strategies*. Springer, Berlin, pp 261–316.

Hernandez, J., Jimenez, A. and Millineaus, P., Sevilla, F. (2000). Tolerance to pea (*Pisum sativum* L.) to long-term salt stress is associated with induction of antioxidant defenses. *Plant Cell Environment,* 23:853–862.

Hernandez, J. A., del Rıo, L. A. and Sevilla, F. (2006). Salt stress-induced changes in superoxide dismutase isozymes in leaves and mesophyll protoplasts from *Vigna unguiculata* (L.) Walp. *New Phytology,* 126: 37-44.

Holzmeister, C., Gaupels, F., Geerlof, A., Sarioglu, H., Sattler, M., Durner, J. and Lindermayr, C. (2015). Differential inhibition of *Arabidopsis* superoxide dismutases by peroxynitrite-mediated tyrosine nitration. *Journal of Experimental Botany,* 66:989–999.

Huang, M. and Guo, Z. (2005). Responses of antioxidative system to chilling in two rice cultivars differing in sensitivity. *Biological Plant,* 49: 81-84.

Inzé, D. and Van Montagu, M. (1995). Oxidative stress in plants. *Current Opinion in Biotechnology,* 6: 153-158.

Ismail, M. and Suroto, A. (2012). Response of biochemical activities on selected rice plants towards ozone stress. *International Conference on Nanotechnology and Chemical Engineering,* December, 2002, Thailand, pp. 21-22.

Janas, K. M., Zielinska-Tomaszewska, J., Rybaczek, D., Maszewski, J., Posmyk, M. M., Amarowicz, R., et al. (2010). The impact of copper ions on growth, lipid peroxidation, and phenolic compound accumulation and localization in lentil (*Lens culinaris Medic.*) seedlings. *Journal of Plant Physiology,* 167: 270-276.

Jia, X., Wang, W. X., Ren, L., Chen, Q. J, Mendu, V. et al. (2009). Differential and dynamic regulation of miR398 in response to ABA and salt stress in *Populus tremula* and *Arabidopsis thaliana*. *Plant Molecular Biology,* 71:51–59.

Jiang, Y. and Huang, B. (2001). Drought and heat stress injury to two cool-season turf grasses in relation to antioxidant metabolism and lipid peroxidation. *Crop Science*, 41: 436-444.

Kampfenkel, K., Van Montagu, M. and Inzé, D. (1995). Effects of iron excess on *Nicotiana plumbaginifolia* plants (implications to oxidative stress). *Plant Physiology*, 107: 725-735.

Kanematsu, S. and Asada, K. (1989). Cu/Zn superoxide dismutase in rice: occurrence of an active monomeric enzyme and two types of isozymes in leaf and non-photosynthetic tissues. *Plant Cell Physiology,* 30:381–391.

Kanematsu, S., Okayasu, M. and Ueno, S. (2013). Atypical cytosol-localized Fe superoxide dismutase in the moss *Pogonatum inflexum*. *Bulletin Minamikyushu Univ*, 3A(23):31.

Kang, G., Wang, C., Sun, G. and Wang, Z. (2003). Salicylic acid changes activities of H_2O_2-metabolizing enzymes and increases the chilling tolerance of banana seedlings. *Environmental and Experimental Botany,* 50: 9-15.

Karlsson, M., Melzer, M., Prokhorenko, I., Johansson, T. and Wingsle, G. (2005). Hydrogen peroxide and expression of hipI-superoxide dismutase are associated with the development of secondary cell walls in *Zinnia elegans*. *Journal of Experimental Botany,* 56:2085–2093.

Karpinska, B., Karlsson, M., Srivastava, M., Stenberg, A., Schrader, J., Sterky, F., Bhalerao, R. and Wingsle, G. (2004). MYB transcription factors are differentially expressed and regulated during secondary vascular tissue development in hybrid aspen. *Plant Molecular Biology,* 56:255–270.

Kasai, T., Suzuki, T., Ogawa, K. O. K., Inagaki, Y., Ichinose, Y., Toyoda, K. and Shiraishi, T. (2006). Pea extracellular Cu/Zn-superoxide dismutase responsive to signal molecules from a fungal pathogen. *Journal of General Plant Pathology*, 72:265–272.

Khan, N. A., Samiullah, Singh, S. and Nazar, R. (2007). Activities of antioxidative enzymes, sulphur assimilation, photosynthetic activity and growth of wheat (*Triticum aestivum*) cultivars differing in yield potential under cadmium stress. *Journal of Agronomy and Crop Science,* 193:435–444.

Kim, H. J., Kato, N., Kim, S. and Triplett, B. (2008). Cu/Zn superoxide dismutases in developing cotton fibers: evidence for an extracellular form. *Planta,* 228:281–292.

Kotak, S., Larkindale, J., Lee, U., von Koskull-Doring, P., Vierling, E. and Scharf, K. D. (2007). Complexity of the heat stress response in plants. *Current Opinion in Plant Biology,* 10: 310-316.

Kranner, I. and Colville, L. (2011). Metals and seeds: biochemical and molecular implications and their significance for seed germination. *Environmental and Experimental Botanique,* 72: 93-105.

Kumar, D., Yusuf, M. A., Singh, P., Sardar, M. and Sarin, N. B. (2013). Modulation of antioxidant machinery in α-tocopherol-enriched transgenic *Brassica juncea* plants tolerant to abiotic stress conditions. *Protoplasma,* 250: 1079-1089.

Kumar, P., Kumar, R. T. and Sharma, P. N. (2007). Excess nickel-induced changes in antioxidative processes in maize leaves. *Journal of Plant Nutrition and Soil Science,* 170: 796-802.

Kwiatowski, J. and Kaniuga, Z. (1986). Isolation and characterization of cytosolic and chloroplastic isozymes of Cu-Zn superoxide dismutase from tomato leaves and their relationships to other Cu-Zn superoxide dismutases. *Biochimica and Biophysica Acta,* 874:99–115.

Kwon, S., Jeong, Y., Lee, H., Kim, J., Cho, K., Allen, R. and Kwak, S. (2002). Enhanced tolerances of transgenic tobacco plants expressing both superoxide dismutase and ascorbate peroxidase in chloroplasts against methyl viologen mediated oxidative stress. *Plant Cell Environment,* 25:873–882.

Lee, D. H. and Lee, C. B. (2000). Chilling stress-induced changes of antioxidant enzymes in the leaves of cucumber: in gel enzyme activity assays. *Plant Science,* 159: 75-85.

Lee, Y. P., Ahmad, R., Lee, H. S., Kwak, S. S., Shafqat, M. N. and Kwon, S. Y. (2013). Improved tolerance of Cu/Zn superoxide dismutase and ascorbate peroxidase expressing transgenic tobacco seeds and seedlings against multiple abiotic stresses. *International Journal of Agriculture and Biology,* 15: 725-730.

Luo, X., Wu, J., Li, Y., Nan, Z., Guo, X., Wang, Y., Zhang, A., Wang, Z., Xia, G. and Tian, Y. (2013). Synergistic effects of GhSOD1 and GhCAT1 overexpression in cotton chloroplasts on enhancing tolerance to methyl viologen and salt stresses. *PLoS One*, 8(1):e54002.

Mackerness, S. H. A. (2000). Plant responses to ultraviolet-B (UV-B: 280_320 nm) stress: what are the key regulators? *Plant Growth Regulation,* 32: 27-39.

Madamanchi, N. R. and Alscher, R. G. (1991). Metabolic bases for differences in sensitivity of two pea cultivars to sulfur dioxide. *Plant Physiology,* 97: 88-93.

Mahanty, S., Kaul, T., Pandey, P., Reddy, R. A., Mallikarjuna, G., Reddy, C. S., et al. (2012). Biochemical and molecular analyses of copper_zinc superoxide dismutase from a C4 *Pennisetum glaucum* reveals an adaptive role in response to oxidative stress. *Gene*, 505: 309-317.

Mann, T. and Keilin, D. (1938). Haemocuprein and hepatocuprein, copper protein compounds of blood and liver in mammals. *Proceeding of the Royal Society*, London 126:303–315.

McCord, J. and Fridovich, I. (1969). Superoxide dismutase. An enzymic function for erythrocuprcin (hemocuprein). *Journal of Biochemistry,* 244:6049–6055.

Marino, M., Galvano, M., Cambria, A., Polticelli, F. and Desideri, A. (1995). Modeling the three-dimensional structure and the electrostatic potential field of two Cu, Zn superoxide dismutase variants from tomato leaves. *Protein Engineering,* 8:551–556.

Martinez, C. A, Loureiro, M. E, Oliva, M. A. and Maestri, M. (2001). Differential responses of superoxide dismutase in freezing resistant *Solanum curtilobum* and freezing sensitive *Solanum tuberosum* subjected to oxidative and water stress. *Plant Science,* 160:505–515.

McKersie, B. D., Chen, Y., de Beus, M., Bowley, S. R., Bowler, C. et al. (1993). Superoxide dismutase enhances tolerance of freezing stress in transgenic alfalfa (*Medicago sativa* L.). *Plant Physiology*, 103:1155–1163.

McKersie, B. D., Bowley, S. R. and Jones, K. S. (1999). Winter survival of transgenic alfalfa overexpressing superoxide dismutase. *Plant Physiology*, 119: 839–848.

Milone, M. T., Sgherri, C., Clijsters, H. and Navari-Izzo, F. (2003). Antioxidative responses of wheat treated with realistic concentration of cadmium. *Environmental and Experimental Botany,* 50: 265-276.

Mishra, S., Srivastava, S., Tripathi, R.D., Govindrajan, R., Kuriakose, S.V. and Prasad, M. N. V. (2006). Phytochelatin synthesis and response of antioxidants during cadmium stress in *Bacopa monnieri* L. *Plant Physiology and Biochemistry,* 44:25–37.

Mishra, P., Bhoomika, K. and Dubey, R. S. (2013). Differential responses of antioxidative defense system to prolonged salinity stress in salt-tolerant and salt-sensitive Indica rice (*Oryza sativa* L.) seedlings. *Protoplasma,* 250:3–19.

Mittler, R. (2002). Oxidative stress, antioxidants and stress tolerance. *Trends in Plant Science,* 7: 405-410.

Mittler, R. (2006). Abiotic stress, the field environment and stress combination. *Trends Plant Science,* 11: 15-19.

Mobin, M. and Khan, N. A. (2007). Photosynthetic activity, pigment composition and antioxidative response of two mustard (*Brassica juncea*) cultivars differing in photosynthetic capacity subjected to cadmium stress. *Journal of Plant Physiology*, 164: 601–610.

Mou, D., Yao, Y., Yang, Y., Zhang, Y., Tian, C. and Achal, V. (2011). Plant high tolerance to excess manganese related with root growth, manganese distribution and antioxidative enzyme activity in three grape cultivars. *Ecotoxicology and Environmental Safety,* 74: 776-786.

Mutlu, S., Karadagoglu, O., Atici, O. and Nalbantoglu, B. (2013). Protective role of salicylic acid applied before cold stress on antioxidative system and protein patterns in barley apoplast. *Biologia Plantarum,* 57: 507-513.

Mylona, P. V. and Polidoros, A. N. (2010). ROS regulation of antioxidant genes. In: Gupta SD (ed.) Reactive oxygen species and antioxidants in higher plants. *Science Publishers*, pp. 101–127.

Myouga, F., Hosoda, C., Umezawa, T., Iizumi, H., Kuromori, T. et al. (2008). A heterocomplex of iron superoxide dismutases defends chloroplast nucleoids against oxidative stress and is essential for chloroplast development in *Arabidopsis*. *Plant Cell Online*, 20:3148–3162.

Negi, N. P., Shrivastava, D. C., Sharma, V. and Sarin, N.B. (2015). Overexpression of CuZnSOD from *Arachis hypogaea* alleviates salinity and drought stress in tobacco. *Plant Cell Report*, 34(7):1109-1126.

Ozdener, Y. and Aydin, B. K. (2010). The effect of zinc on the growth and physiological and biochemical parameters in seedlings of *Eruca sativa* (L.) (Rocket). *Acta Physiologia Plantarum,* 32: 469-476.

Potters, G., Pasternak, T. P., Guisez, Y., Palme, K. J. and Jansen, M. A. K. (2007). Stress-induced morphogenic responses: growing out of trouble? *Trends Plant Science*, 12: 98-105.

Prasad, K. V. S. K., Paradha, S. P. and Sharmila, P. (1999). Concerted action of antioxidant enzymes and curtailed growth under zinc toxicity in *Brassica juncea*. *Environmental and Experimental Botany,* 42: 1-10.

Prasad, T. K., Anderson, M. D., Martin, B. A. and Stewart, C. R. (1994). Evidence for chilling-induced oxidative stress in maize seedlings and a regulatory role for hydrogen peroxide. *Plant Cell*, 6: 65-74.

Rao, M. V., Paliyath, G. and Ormrod, D. P. (1996). Ultraviolet-B-and ozone-induced biochemical changes in antioxidant enzymes of *Arabidopsis thaliana*. *Plant Physiology*, 110: 125-136.

Rasool, S., Ahmad, A., Siddiqi, T. O. and Ahmad, P. (2013). Changes in growth, lipid peroxidation and some key antioxidant enzymes in chickpea genotypes under salt stress. *Acta Physiologia Plantarum*, 35: 1039-1050.

Rizhsky, L., Liang, H. and Mittler, R. (2003). The water-water cycle is essential for chloroplast protection in the absence of stress. *The Journal of Biological Chemistry,* 278: 38921–38925.

Rivero, R. M., Ruiz, J. M., Garcia, P. C., Lopez-Lefebre, L. R., Sanchez, E. and Romero, L. (2001). Resistance to cold and heat stress: accumulation of phenolic compounds in tomato and water melon plants. *Plant Science,* 160: 315-321.

Rout, G. R. and Das, P. (2003). Effect of metal toxicity on plant growth and metabolism: I. Zinc. *Agronomie,* 23: 3-11.

Ruzsa, S. M., Mylona, P. and Scandalios, J. G. (1999). Differential responses of antioxidant genes in maize leaves exposed to ozone. *Redox Report,* 4: 95–103.

Sairam, R. K., Srivastava, G. C., Agarwal, S. and Meena, R.C. (2005). Differences in antioxidant activity in response to salinity stress in tolerant and susceptible wheat genotypes. *Biologia Plantarum,* 49:85–91.

Schickler, H. and Caspi, H. (2002). Response of antioxidative enzymes to nickel and cadmium stress in hyperaccumulator plants of the genus *Alyssum*. *Physiologia Plantarum,* 105, 39-44.

Schinkel, H., Hertzberg, M. and Wingsle, G. (2001). A small family of novel Cu/Zn-superoxide dismutases with high isoelectric points in hybrid aspen. *Planta,* 213:272–279.

Schutzendubel, A. and Polle, A. (2002). Plant responses to abiotic stresses: heavy metal-induced oxidative stress and protection by mycorrhization. *Journal of Experimental Botany,* 53: 1351-1365.

Sen, A. and Alikamanoglu, S. (2013). Antioxidant enzyme activities, malondialdehyde, and total phenolic content of PEG-induced hyperhydric leaves in sugar beet tissue culture. *In Vitro Cellular and Developmental Biology-Plant,* 49: 396-404.

Shafi, A., Dogra, V., Gill, T., Ahuja, P. S. and Sreenivasulu, Y. (2014). Simultaneous over-expression of PaSOD and RaAPX in transgenic *Arabidopsis thaliana* confers cold stress tolerance through increase in vascular lignifications. *PLoS One,* 9(10):e110302.

Shainberg, O., Rubin, B., Rabinowitch, H. D., Libal, Y. and Tel-Or, E. (2000). Acclimation of beans to oxidative stress by treatment with sub-lethal iron levels. *Journal of Plant Physiology,* 157: 93-99.

Shalata, A., Mittova, V., Volokita, M., Guy, M. and Tal, M. (2001). Response of the cultivated tomato and its wild salt-tolerant relative *Lycopersicon pennellii* to salt-dependent oxidative stress: the root antioxidative system. *Physiologia Plantarum*, 112:487–494.

Sharma, P. and Dubey, R. S. (2005). Drought induces oxidative stress and enhances the activities of antioxidant enzymes in growing rice seedlings. *Plant Growth Regulation,* 46: 209-221.

Sharma, P., Jha, A. B., Dubey, R. S., Pessarakli, M. (2012). Reactive oxygen species, oxidative damage, and antioxidative defense mechanism in plants under stressful conditions. *Journal of Botany*, 1-26.

Sheokand, S., Kumari, A. and Sawhney, V. (2008). Effect of nitric oxide and putrescine on antioxidative responses under NaCl stress in chickpea plants. *Physiology and Molecular Biology of Plants*, 14, 355-362.

Silva, E. N., Vieira, S. A., Ribeiro, R. V., Ponte, L. F., Ferreira-Silva, S. L. and Silveira, J. A. (2013). Contrasting physiological responses of *Jatropha curcas* plants to single and combined stresses of salinity and heat. *Journal of Plant Growth Regulation,* 32: 159-169.

Singh, S., Khan, N. A., Nazar, R. and Anjum, N. A. (2008). Photosynthetic traits and activities of antioxidant enzymes in blackgram (*Vigna mungo* L. Hepper) under cadmium stress. *American Journal of Plant Physiology,* 3:25–32.

Sinha, S. and Saxena, R. (2006). Effect of iron on lipid peroxidation, and enzymatic and non-enzymatic antioxidants and bacoside-A content in medicinal plant *Bacopa monnieri* L. *Chemosphere*, 62: 1340-1350.

Srivastava, S. and Dubey, R. S. (2011). Manganese-excess induces oxidative stress, lowers the pool of antioxidants and elevates activities of key antioxidative enzymes in rice seedlings. *Plant Growth Regulation,* 64: 1-16.

Sunkar, R., Kapoor, A. and Zhu, J. K. (2006) Post-transcriptional induction of two Cu/Zn superoxide dismutase genes in *Arabidopsis* is mediated by down-regulation of miR398 and important for oxidative stress tolerance. *Plant Cell,* 18:2051–2065.

Szepesi, A., Poor, P., Gémes, K., Horvath, E. and Tari, I. (2008). Influence of exogenous salicylic acid on antioxidant enzyme activities in the roots of salt stressed tomato plants. *Acta Biologica Szegediensis,* 52: 199-200.

Szollosi, R., Sz. Varga, I., Erdei, L. and Mihalik, E. (2009). Cadmium-induced oxidative stress and antioxidative mechanisms in germinating Indian mustard (*Brassica juncea* L.) seeds. *Ecotoxicology Environmental Safety,* 72: 1337-1342.

Szollosi, R., Kalman, E., Medvegy, A., Peto, A. and Sz. Varga, I. (2011). Studies on oxidative stress caused by Cu and Zn excess in germinating seeds of Indian mustard (*Brassica juncea* L.). *Acta Biologica Szegediensis,* 55: 175-178.

Szollosi, R., Kalman, E., Medvegy, A. and Sz. Varga, I. (2012). Survival to early toxic copper excess: biochemical and anatomical changes during germination of Indian Mustard. *European Chemical Bulletin* 1, 466-469.

Tang, L., Kwon, S. Y., Kim, S. H., Kim, J. S., Choi, J. S., Cho, K. Y., Sung, C. K., Kwak, S. S. and Lee, H. S. (2006). Enhanced tolerance of transgenic potato plants expressing both superoxide dismutase and ascorbate peroxidase in chloroplasts against oxidative stress and high temperature. *Plant Cell Report,* 25(12):1380-1386.

Tanyolac, D., Ekmekc, Y. and Unalan, S. (2007). Changes in photo chemical and antioxidant enzyme activities in maize (*Zea mays* L.) leaves exposed to excess copper. *Chemosphere*, 67:89–98.

Tari, I., Camen, D., Coradini, G., Csiszar, J., Fediuc, E., Gémes, K., et al. (2008). Changes in chlorophyll fluorescence parameters and oxidative stress responses of bush bean genotypes for selecting contrasting acclimation strategies under water stress. *Acta Biologica Hungarica,* 59: 335-345.

Tepperman, J. M. and Dunsmuir, P. (1990). Transformed plants with elevated levels of chloroplastic SOD are not more resistant to superoxide toxity. *Plant Molecular Biology*, 14:501–511.

Tertivanidis, K., Goudoula, C., Vasilikiotis, C., Hassiotou, E., Perl-Treves, R., Tsaftaris, A. (2004). Superoxide dismutase transgenes in sugar beets confer resistance to oxidative agents and the Fungus *C. beticola. Transgenic Research,* 13:225–233.

Tuteja, N. and Gill, S. S. (2013). *Crop improvement under adverse conditions*, 1st ed. Springer, New York.

Tuteja, N., Mishra, P., Yadav, S., Tajrishi, M., Baral, S. and Sabat, S. C. (2015). Heterologous expression and biochemical characterization of a highly active and stable chloroplastic CuZn-superoxide dismutase from *Pisum sativum. BMC Biotechnology,* 15:3.

Unyayar, S., Keles, Y. and Cekic, F. O. (2005). The antioxidative response of two tomato species with different drought tolerances as a result of drought and cadmium stress combinations. *Plant Soil Environment*, 51, 57-64.

Van Camp, W., Capiau, K., Van Montagu, M., Inzé, D. and Slooten, L. (1996). Enhancement of oxidative stress tolerance in transgenic tobacco plants overexpressing Fe-superoxide dismutase in chloroplasts. *Plant Physiology*, 112:1703–1714.

Wang, F. Z., Wang, Q. B., Know, S.Y., Kwak, S. S. and Su W. A. (2005) Enhanced drought tolerance of transgenic rice plants expressing a pea manganese superoxide dismutase. *Journal of Plant Physiology,* 162:465–472.

Wang, C., Zhang, S. H., Wang, P. F., Hou, J., Zhang, W. J., Li, W., et al. (2009). The effect of excess Zn on mineral nutrition and antioxidative response in rape seed seedlings. *Chemosphere,* 75: 1468-1476.

Willekens, H., Van Camp, W., Van Montagu, M., Inzé, D., Langebartels, C. and Sandermann Jr. H. (1994). Ozone, sulfur dioxide, and ultraviolet B have similar effects on mRNA accumulation of antioxidant genes in *Nicotiana plumbaginifolia* L. *Plant Physiology,* 106: 1007-1014.

Wu, F. B., Dong, J., Jia, G. X., Zheng, S. J. and Zhang, G. P. (2006). Genotypic differences in the responses of seedling growth and Cd toxicity in rice (*Oryza sativa* L.). *Agriculture Science China,* 5:68–76.

Xu, S. C., Li, Y. P., Hu, J., Guan, Y. J., Ma, W. G., Zheng, Y. Y., et al. (2010). Responses of antioxidant enzymes to chilling stress in tobacco seedlings. *Agriculture Science China*, 9: 1594-1601.

Xu, J., Yang, J., Duan, X., Jiang, Y. and Zhang, P. (2014). Increased expression of native cytosolic Cu/Zn superoxide dismutase and ascorbate peroxidase improves tolerance to oxidative and chilling stresses in cassava (*Manihot esculenta Crantz*). *BMC Plant Biology,* 14 (208).

Xue, S. G., Chen, Y. X., Reeves, R. D., Baker, A. J., Lin, Q. and Fernando, D. R. (2004). Manganese uptake and accumulation by the hyper-accumulator plant *Phytolacca acinosa* Roxb. (Phytolaccaceae). *Environmental Pollution*, 131: 393-399.

Yadav, S. K. (2010). Heavy metals toxicity in plants: an overview on the role of glutathione and phytochelatins in heavy metal stress tolerance of plants. *South Africa of Journal of Botanic,* 76: 167-179.

Yousuf, P. Y., Hakeem, K. U. R., Chandna, R. and Ahmad, P. (2012). Role of glutathione reductase in plant abiotic stress. In: Ahmad, P., Prasad, M.N.V. (Eds.), *Abiotic Stress Responses in Plants: Metabolism, Productivity and Sustainability*. Springer, New York, pp. 149-158.

Zagorchev, L., Seal, C. E., Kranner, I. and Odjakova, M. (2013). A central role for thiols in plant tolerance to abiotic stress. *International Journal of Molecular Science,* 14: 7405-7432.

Zhang, J. and Kirkham, M. (1995). Water relations of water-stressed, split-root C4 (*Sorghum bicolor;* Poaceae) and C3 (*Helianthus annuus*; Asteraceae) plants. *American Journal of Botanic,* 82:1220–1229.

Zhang, H., Xia, Y., Wang, G. and Shen, Z. (2008). Excess copper induces accumulation of hydrogen peroxide and increases lipid peroxidation and total activity of copper_zinc superoxide dismutase in roots of *Elsholtzia haichowensis*. *Planta,* 227: 465-475.

Zsigmond, L., Szepesi, A., Tari, I., Rigo, G., Kiraly, A. and Szabados, L. (2012). Overexpression of the mitochondrial PPR40 gene improves salt tolerance in *Arabidopsis*. *Plant Science,* 182: 87-93.

BIOGRAPHICAL SKETCH

Faiçal Brini

Affiliation: Biotechnology and Plant Improvement Laboratory, Centre of Biotechnology of Sfax-Tunisia

Education:

- 2001-2006: PhD, Plant Molecular Biology, National School of Engineer of Sfax - Tunisia
- 1998-2000: Master of Science, National School of Engineer of Sfax - Tunisia.
- 1994-1998: Maitrise in Science, University of Science of Sfax - Tunisia

Business Address: Centre of Biotechnology of Sfax (CBS); Road sidi Mansour km 6, B.P.'1177' 3018 Sfax-Tunisia.

Research and Professional Experience:

- January 2015 at present: Associate Professor and Head of the Laboratory of Biotechnology and Plant Improvement, Centre of Biotechnology of Sfax- Tunisia.
- January 2013- December 2014: Associate Professor and Head of the Plant Team, Laboratory of Plant Protection and Improvement, Centre of Biotechnology of Sfax- Tunisia.
- October 2006 – December 2012: Associate Professor and member of the Laboratory of Plant Protection and Improvement, Centre of Biotechnology of Sfax- Tunisia.
- October 2004 – 2006: Assistant Professor, Faculty of Science of Gabès-Tunisia.

Professional Appointments:

Plant Molecular Biology; Plant Physiology; Biochemistry; Microbiology and Plant Transformation.

Honors:

Price of the best poster exposed at the university fair in 2015.

Publications from the Last 3 Years:

Tounsi S, Feki K, Hmidi D, masmoudi Kh, Brini F (2017) Salt stress reveals differential physiological, biochemical and molecular responses in *T. monococcum* and *T. durum* wheat genotypes. *Physiology and Molecular Biology of Plants*, DOI 10.1007/s12298-017-0457-4.

Dabbous A, Ben Saad R, Brini F, Farhat-Khemekhem A, Zorrig W, Abdely Ch, Ben Hamed K (2017) Over-expression of a subunit E1 of a vacuolar H^+-ATPase gene (*Lm VHA-E1*) cloned from the halophyte *Lobularia maritima* improves the tolerance of *Arabidopsis thaliana* to salt and osmotic stresses. *Environmental and Experimental Botany* 137: 128-141.

Said MN, Mergby Dh, Brini F (2017) Identification and expression analysis of the NAC transcription factor family in durum wheat (Triticum turgidum L. ssp. durum). *Plant Physiology and Biochemistry*, 112: 117-128.

Feki K, Tounsi S, Masmoudi K, Brini F (2016) The durum wheat plasma membrane Na^+/H^+ antiporter SOS1 is involved in oxidative stress response. *Protoplasma*, DOI 10.1007/s00709-016-1066-8.

Drira M, Hanin M, Masmoudi K, Brini F (2016)Comparison of full-length and conserved segments of wheat dehydrin DHN-5 overexpressed in Arabidopsis thaliana showed different responses to abiotic and biotic stress. *Functional Plant Biology*, 43 (11), 1048-1060.

Brini F (2016) Genetic Transformation of Wheat: Current Status and Futures Challenges. Book entitled "*Agricultural Research Updates*." ISBN: 978-1-63485-648-5, edited by Prathamesh Gorawala and Srushti Mandhatri. NOVA Science Publishers, vol 13, Chapter 6, pp 105-131.

Benderradji L, Kellou K, Ghadbane M, Salmi M, Saibi W, Benmahammed A, Bouzerzour H, Brini F (2016)Effects of Plant Growth Promoting Rhizobacteria (PGPR) on *in vitro* bread wheat (*Triticum aestivum* L.)

growth parameters and biological control mechanisms. *Advances in Microbiology*, 6: 677-690.

Tounsi S, Ben Amar S, Masmoudi K, Sentenac H, Brini F, Véry AA (2016)Characterisation of two HKT1;4 Transporters from *Triticum monococcum* to elucidate the Determinants of the Wheat Salt Tolerance *Nax1* QTL. *Plant Cell Physiology*.57 (10), 2047-2057, doi: 10.1093/pcp/pcw123.

Boubakri H, Gargouri M, Mliki A, Brini F, Chong J, Jbara M (2016) Vitamins for enhancing plant resistance. *Planta*, DOI 10.1007/s00425-016-2552-0.

Benderradji L, Hadji N, Kellou K, Benniou R, Brini F (2016) Effet du NaCl et du PEG sur le comportement morpho-physiologique et biochimique des variétés de blé dur et tendre cultivées in vitro en milieu hydroponique. *Revue Agriculture*, Université Abbas Sétif 1. 1:278-286.

Saibi W, Zouari N, Masmoudi K, Brini F (2016) Role of the durum wheat dehydrin in the function of proteases conferring salinity tolerance in Arabidopsis thaliana transgenic lines. *International Journal of Biological Macromolecules*. 85: 311-316.

Feki K, Farhat-Khemakhem A, Kammoun Y, Saibi W, Gargouri A, Brini F (2016) Responses of transgenic Arabidopsis plants and recombinant yeast cells expressing a novel durum wheat manganese superoxide dismutase TdMnSOD to various abiotic stresses. *Journal of Plant Physiology*. 198: 56-68.

Feki K, Brini F (2016) Role of proteins in alleviating drought stress in plants. Book entitled "*Water Stress and Crop Plants: A Sustainable Approach*" edited by Parvaiz Ahmed. Chapter 12, pp 165-176.

Feki K, Brini F, Ben Amar S, Saibi W, Masmoudi K (2015) Comparative functional analysis of two wheat Na^+/H^+ antiporter SOS1 promoters in Arabidopsis thaliana under various stress conditions. *Journal of Applied genetics*. 56, (11), 15-26.

Drira M, Saibi W, Amara I, Masmoudi K, Hanin M, Brini F (2015) Wheat Dehydrin K-Segments Ensure Bacterial Stress Tolerance, Antiaggregation and Antimicrobial Effects. *Applied Biochemistry and Biotechnology*. 175 (7): 3310-3321.

Saibi W, Drira M, Yakoubi I, Feki K, Brini F (2015) Empiric, structural and in silico findings give birth to plausible explanations for the multifunctionality of the wheat dehydrin (DHN-5). *Acta Physiologia Plantarum*. 37:52. DOI: 10.1007/s11738-015-1798-7.

Safi H, Saibi W, Alaoui MM, Hmyene A, Masmoudi K, Hanin M, Brini F (2015) A wheat lipid transfer protein (TdLTP4) promotes tolerance to abiotic and biotic stress in Arabidopsis thaliana. *Plant Physiology and Biochemistry*. 89: 64-75.

Saibi W, Feki K, Yacoubi I, Brini F (2015) Bridging Between Proline Structure, Functions, Metabolism, and Involvement in Organism Physiology. *Applied Biochemistry and Biotechnology*. 176: 2107-2119.

Saibi W, Feki K, Ben Mahmoud R, Brini F (2015) Durum wheat dehydrin (DHN-5) confers salinity tolerance to transgenic Arabidopsis plants through the regulation of proline metabolism and ROS scavenging system. *Planta* 242 (5): 1187-94.

Feki K, Yosra Kamoun Y, Ben Mahmoud R, Farhat-Khemakhem A, Gargouri A, Brini F (2015) Multiple abiotic stress tolerance of the transformants yeast cells and the transgenic Arabidopsis plants expressing a novel durum wheat catalase. *Plant Physiology and Biochemistry*. 97: 420-431.

In: Superoxide Dismutase
Editor: Sergei Magliozzi

ISBN: 978-1-53613-080-5

Chapter 4

SUPEROXIDE DISMUTASE AND SENSORINEURAL HEARING LOSS

V. Corazzi, C. Bianchini, C. Aimoni and A. Ciorba[*]
ENT and Audiology Department,
University Hospital of Ferrara, Italy

ABSTRACT

Superoxide dismutase (SOD) is an antioxidant agent that can be found in almost every cell exposed to oxygen. It is responsible for the dismutation of superoxide anions resulting from cellular metabolism into oxygen and hydrogen peroxide. In particular, superoxide dismutase (SOD), in its cytoplasmic isoform, contains copper (Cu) and zinc (Zn), and can be found in cochlear tissue (stria vascularis, spiral ligament, spiral ganglion neurons and organ of Corti). The antioxidant function of SOD is important for protection of the inner ear cells from toxic substances, such as reactive oxygen species (ROS). The role of SOD has been investigated in noise-induced hearing loss and in presbycusis, whereas elevated concentrations of ROS are responsible for inner ear damage. In the last decades, many authors have tested the protective role of SOD in the cochlea of animal models. Recently, the role of SOD gene

[*] Corresponding Author Email: andrea.ciorba@unife.it.

polymorphisms in the susceptibility of sudden sensorineural hearing loss has been investigated as well.

The aim of this chapter is to review the role of SOD in the cochlear metabolism and its involvement in the pathogenesis of noise induced hearing loss, age related hearing loss and sudden sensorineural hearing loss.

Keywords: sensorineural hearing loss, superoxide dismutase, reactive oxygen species, oxidative stress, noise-induced hearing loss, presbycusis

INTRODUCTION

The oxygen metabolism represents the fundamental engine for aerobic cells, which is necessary for the cellular production of energy. Reactive oxygen species (ROS), such as hydrogen peroxide, superoxide and hydroxyl radical, are chemically reactive elements obtained as a normal product of the oxygen metabolism and play an important role in cellular homeostasis and signalling [1]. In a standard cellular environment, ROS oxidant activity is counterbalanced by an antioxidant system, which represents the cellular antiradical defence, including some molecules such as glutathione peroxidase, catalase, vitamin A, C and E and superoxide dismutase (SOD) [2].

Several conditions, such as exposition to ionizing radiation and inflammatory processes, may determine an augmented production of ROS, configuring a so-called 'oxidative stress' status. In such circumstances, a loss of balance between oxidant and antioxidant agents occurs and the excessive ROS concentration can cause cellular damage. The mechanisms, through which incremented ROS levels may cause cellular injury, are well established: lipid perixodation, protein oxidation and DNA damage. These mechanisms are also able to induce cellular death through necrosis or apoptosis [1].

Oxidative stress is known to be involved in the normal ageing process [3]; also, it is known to play a pivotal role in the pathogenesis of different diseases, such as carcinogenesis [4], ischaemia [5] and neurodegenerative disorders, including Alzheimer disease [6]. Furthermore, it has been reported that oxidative stress may be linked to hearing loss [7]: incremented ROS levels, among inner ear cells, have been reported in several conditions such as presbycusis [8], ototoxicity (i.e., cisplatin, aminoglycosides) [9] and noise exposure [10]. Inner ear hair cells are particularly susceptible to variation of cochlear blood flow and to oxygen levels [11]. Moreover, the cochlear lateral wall (stria vascularis and spiral ligament), involved in the preservation of the endolymphatic homeostasis and the endocochlear potential, is particular vulnerable to ischemic conditions [12].

It has been described that in case of cochlear ischemic injury, an increased ROS production may determine inner ear cells damage through two mechanisms: (i) a direct cell metabolic impairment and (ii) an endothelial dysfunction [13]. Some authors [14,15,16,17] have also linked oxidative stress to the onset of sensorineural hearing loss (SNHL), describing a correlation between ROS-induced endothelial dysfunction and inner ear disease.

In this scenario, SOD represents an important defence system against oxidative stress in the inner ear.

SUPEROXIDE DISMUTASE AND THE INNER EAR

SOD is a cellular oxidoreductase enzyme involved in the dismutation of the superoxide into hydrogen peroxide. SOD is a crucial cellular antioxidant element, which can be found in almost every cell exposed to oxygen [18]. In mammals there are three isoforms of SOD: they differ mainly in the ion included in the active site (copper, iron, zinc, nickel and manganese) and in their distribution: SOD1 is located in the cytoplasm, SOD2 in mitochondria, while SOD3 is an extracellular form [18].

During the last decades, many authors have already demonstrated the crucial importance of the oxidoreductive function of SOD with animal models. Mice lacking SOD activity develop several pathologies, such as hepatocellular carcinoma [19], early cataract [20], and also hearing loss

[21]. Furthermore, other authors have demonstrated the protective role of SOD towards outer hair cells in vitro [22] and in the mice cochlea [23].

SOD1, the cytoplasmic isoform, has been found prevalently in cochlear tissue (stria vascularis, spiral ligament, spiral ganglion neurons and organ of Corti), with a minor concentration also located in the mitochondria intramembrane space [24,25]. Stria vascularis retains high SOD concentration, as it is involved in the maintenance of the endocochlear potential due to its central role in ionic transportation; in addition, this process requires a continue active aerobic oxidation in order to obtain ATP [12]. High SOD concentration in stria vascularis plays a relevant role in detoxification from oxygen free radicals [25]. In spiral ganglion neurons and in the organ of Corti, SOD has been reported to protect directly neural and hair cells from ROS damages as well [21].

Superoxide Dismutase and Noise-Induced Hearing Loss

The pathogenic role of oxidative stress in noise-induced hearing loss has been already described [26]; increased levels of ROS after noise exposure have been linked to hair cells damage [27]. Recently, Fetoni et al., [10] demonstrated the morphological and functional alterations of hair cells and spiral ganglion neurons after a repeated noise exposure in a rat model: they described a cochlear damage, at the medial and basal parts, with a consequent hearing loss as showed by an increased ABR thresholds. This phenomenon was due to an oxidative imbalance, described by the Author as an increase in the cellular production of superoxide and the consequent lipid peroxidation. They also observed an amendment of cochlear injuries and of the auditory thresholds, following the administration of an antioxidant treatment to reverse the noise-induced redox unbalance.

In 1999, Ohlemiller et al., demonstrated an augmented susceptibility to noise-induced hearing impairment in a group of mice lacking SOD1 [28], highlighting the defensive significance of intracellular controlled levels of SOD.

Liu et al., [29] explored the relationship between SOD1 polymorphisms and the susceptibility to noise-induced hearing loss, describing a significant difference between noise-resistant and noise-sensitive workers. In particular, they suggested that rs2070424 and rs10432782 SOD1 polymorphisms might be responsible for an increased susceptibility to noise-induced hearing impairment. A more recent meta-analysis performed by Wang et al., [30] examined the same relationship, considering a specific SOD2 polymorphism (c47t): in this study, the authors did not find any statistically significant correlation.

Superoxide Dismutase and Presbycusis

Oxidative stress is already known to be involved in the aging process; an unbalanced cochlear redox status also plays an important role in the development of presbycusis. A reduction of SOD enzymes levels (especially SOD2) in aged tissues of rats and mice has been comprehensively described more than 30 years ago [31, 32]. The relationship between oxidative stress and the aging of inner ear was investigated in mice [8]: oxidative stress levels in the aging cochlea increase, while antioxidant factors, such as mitochondrial apoptosis-inducing factor and SOD2, decrease. In particular, it has been found that SOD2 levels are reduced in outer hair cells, supporting cells and spiral ganglion cells. McFadden et al, in 1999, demonstrated that in a group of SOD1 lacking mice, the loss of inner ear outer hair cells was greater than in the control wild type mice group, highlighting that the deficiency of SOD activity enhances the age-related degeneration of neuroepithelium of the inner ear [21]. These results are consistent with the study performed by Keithley et al., in a mice model in 2005 [33].

Superoxide Dismutase and Sudden Sensorineural Hearing Loss

Genetic alterations of SOD enzymes have been associated with several diseases; for example, mutations of SOD1 can cause familial amyotrophic lateral sclerosis, a form of motor neuron disease [34]. Different authors examined the role of SOD gene polymorphisms in hearing loss development [29, 30], which appears to be race-specifically distributed. More recently, Kitoh et al., explored the potential correlation between a particular polymorphism of SOD1 gene and the susceptibility to sudden sensorineural hearing loss (SSNHL) [35].

There are many proposed classical etiological theories of SSNHL, including the vascular hypothesis, based on a hemorrhagic or ischemic insult to cochlea [36]. Considering this possible etiology, ROS may represent an important factor in cochlear damage and sudden hearing loss. The relation between SOD genetic polymorphisms and SSNHL has not been largely investigated yet. Kitoh et al., [35] found in SOD1 rs4998557 a polymorphism likely associated with susceptibility to SSNHL, however no SOD2 polymorphisms were found to be correlated with a predisposition to SSNHL. Earlier, Teranishi et al., [37] explored the SOD2 polymorphism in two groups of patients, one group of patients with SSNHL and another group of patients with Ménière's Disease. They found no statistically significant differences in the distribution of polymorphisms among the two groups. In conclusion, they did not suggest any association between specific genotypes and incremented risk of SSNHL or Ménière's Disease. Nonetheless, they described a particular C allele of SOD2 (rs4880) to be associated with progression of the hearing impairment in Ménière patients.

Conclusion

SOD represents an important protective antioxidant system of eukaryotic cells. In the cochlea, SOD seems to play a crucial role in protecting auditory neuroepithelium against toxic substances, such as ROS.

A possible 'therapeutic – protective' role of increased SOD levels has been described. Nonetheless, it has been reported that an over expression of SOD1 in mice does not protect inner ear cells from age-related or noise induced damage [33,38]; however, some authors demonstrated a potential protection provided by SOD against ototoxicity induced by antibiotic drug administration (kanamycin) in mice [23] and in guinea pigs [39].

Recently, Serra et al., [40] proposed a brand-new application of SOD as a nutraceutical treatment in patients with obstructive sleep apnea-hypopnea syndrome (OSAHS), a condition known to be associated with oxidative unbalance. However, a similar approach to inner ear disease treatment has not been previously suggested to our knowledge.

So far, the role of SOD in the pathogenesis of inner ear diseases remains an unresolved issue. Since SOD is involved in hearing function and protection, it could represent a possible therapeutic target for certain inner ear diseases.

REFERENCES

[1] Slimen, I. B.; Najar, T.; Ghram, A.; Dabbebi, H., Ben Mrad, M.; Abdrabbah, M. Reactive oxygen species, heat stress and oxidative-induced mitochondrial damage. A review. *Int. J. Hyperthermia*, 2014; 30(7):513-23.

[2] Vives-Bauza, C.; Starkov, A.; Garcia-Arumi, E. Measurements of the antioxidant enzyme activities of superoxide dismutase, catalase, and glutathione peroxidase. *Methods Cell. Biol.*, 2007; 80:379-93.

[3] Kong, Y.; Trabucco, S. E.; Zhang, H. Oxidative stress, mitochondrial dysfunction and the mitochondria theory of aging. *Interdiscip. Top Gerontol.* 2014;39:86-107.

[4] Lee, J. C.; Son, Y. O.; Pratheeshkumar, P.; Shi, X. Oxidative stress and metal carcinogenesis. *Free Radic. Biol. Med.*, 2012; 53(4):742-57.

[5] Rodrigo, R.; Fernández-Gajardo, R.; Gutiérrez, R.; Matamala, J. M.; Carrasco, R.; Miranda-Merchak, A.; Feuerhake W. Oxidative stress

and pathophysiology of ischemic stroke: novel therapeutic opportunities. *CNS Neurol. Disord. Drug Targets*, 2013; 12(5):698-714.

[6] Swomley, A. M.; Butterfield, D. A. Oxidative stress in Alzheimer disease and mild cognitive impairment: evidence from human data provided by redox proteomics. *Arch. Toxicol.*, 2015; 89(10):1669-80.

[7] Kamogashira, T.; Fujimoto, C.; Yamasoba, T. Reactive oxygen species, apoptosis, and mitochondrial dysfunction in hearing loss. *Biomed. Res. Int.*, 2015; 2015:617207.

[8] Jiang, H.; Talaska, A. E.; Schacht, J.; Sha, S. H. Oxidative imbalance in the aging inner ear. *Neurobiol. Aging.* 2007;28(10):1605-12.

[9] Lanvers-Kaminsky, C.; Zehnhoff-Dinnesen, A. A.; Parfitt, R.; Ciarimboli, G. Drug-induced ototoxicity: Mechanisms, Pharmacogenetics, and protective strategies. *Clin. Pharmacol. Ther.* 2017; 101(4):491-500.

[10] Fetoni, A. R.; De Bartolo, P., Eramo, S. L.; Rolesi, R.; Paciello, F.; Bergamini, C.; Fato, R.; Paludetti, G.; Petrosini, L.; Troiani, D. Noise-induced hearing loss (NIHL) as a target of oxidative stress-mediated damage: cochlear and cortical responses after an increase in antioxidant defense. *J. Neurosci.*, 2013; 33(9):4011-23.

[11] Mudry, A.; Tange, R. A. The vascularization of the human cochlea its historical background. *Acta Oto-Laryngologica*, 2009; 129(Suppl. 561):3-16.

[12] Ciorba, A.; Corazzi, V.; Bianchini, C.; Aimoni, C.; Skarzynski, H.; Skarzynski, P. H.; Hatzopoulos, S. Sudden sensorineural hearing loss: Is there a connection with inner ear electrolytic disorders? A literature review. *Int. J. Immunopathol. Pharmacol.* 2016;29(4):595-602.

[13] Ciorba, A.; Gasparini, P.; Chicca, M.; Pinamonti, S.; Martini, A. Reactive oxygen species in human inner ear perilymph. *Acta Otolaryngol.* 2010;130:240-246.

[14] Guo, Y.; Zhang, C., Du, X.; Nair, U.; Yoo, T. J. Morphological and functional alterations of the cochlea in apolipoprotein E gene deficient mice. *Hear Res.*, 2005; 208:54-67.

[15] Gloddek, B.; Lamm, K.; Arnol, W. Pharmacological influence on inner ear endothelial cells in relation to the pathogenesis of sensorineural hearing loss. *Adv. Otorhinolaryngol.*, 2002; 59:75-83.

[16] Selivanova, O.; Heinrich, U. R.; Brieger, J.; Feltens, R.; Mann, W. Fast alterations of vascular endothelial growth factor (VEGF) expression and that of its receptors (Flt-1, Flk-1 and Neuropilin) in the cochlea of guinea pigs after moderate noise exposure. *Eur. Arch. Otorhinolaryngol.*, 2007; 264: 121-128.

[17] Ciorba, A.; Chicca, A.; Bianchini, C.; Aimoni, C.; Pastore, A. Sensorineural hearing loss and endothelial dysfunction due to oxidative stress: is there a connection? *Int. Adv. Otol.*, 2012; 8(1):16-20.

[18] Fetherol, M. M.; Boyd, S. D.; Winkler, D. D.; Winge, D. R. Oxygen-dependent activation of Cu, Zn-superoxide dismutase-1. *Metallomics*, 2017; 9(8):1047-1059.

[19] Elchuri, S.; Oberley, T. D.; Qi, W.; Eisenstein, R. S.; Jackson Roberts, L.; Van Remmen, H.; Epstein, C. J.; Huang, T. T. CuZnSOD deficiency leads to persistent and widespread oxidative damage and hepatocarcinogenesis later in life. *Oncogene*, 2005; 24(3):367–80.

[20] Behndig, A.; Karlsson, K.; Reaume, A. G.; Sentman, M. L.; Marklund, S. L. In vitro photochemical cataract in mice lacking copper-zinc superoxide dismutase. *Free Radic. Biol. Med.*, 2001; 31(6):738-44.

[21] McFadden, S. L.; Ding, D.; Reaume, A. G.; Flood, D. G.; Salvi, R. J. Age-related cochlear hair cell loss is enhanced in mice lacking copper/zinc superoxide dismutase. *Neurobiol. Aging.* 1999; 20(1):1-8.

[22] Clerici, W. J.; DiMartino, D. L.; Prasad, M. R. Direct effects of reactive oxygen species on cochlear outer hair cell shape in vitro. *Hear. Res.*, 1995; 84(1-2):30-40.

[23] Sha, S. H.; Zajic, G.; Epstein, C. J.; Schacht, J. Overexpression of copper/zinc-superoxide dismutase protects from kanamycin-induced hearing loss. *Audiol. Neurootol.* 2001;6(3):117-23.

[24] Pierson, M. G.; Gray, B. G. Superoxide dismutase activity in the cochlea. *Hear. Res.*, 1982; 6:141-151.

[25] Yao, X.; Rarey, K. E. Detection and regulation of Cu/Zn-SOD and Mn-SOD in rat cochlear tissues. *Hear. Res.*, 1996; 96(1-2):199-203.

[26] Henderson, D.; Bielefeld, E. C.; Harris, K. C.; Hu, B. H. The role of oxidative stress in noise-induced hearing loss. *Ear. Hear.* 2006; 27(1):1-19.

[27] Henderson, D.; McFadden, S. L.; Liu, C. C.; Hight, N.; Zheng, X. Y. The role of antioxidants in protection from impulse noise. *Ann. N Y Acad. Sci.*, 1999; 884: 368-80.

[28] Ohlemiller, K. K.; McFadden, S. L.; Ding, D. L.; Flood, D. G.; Reaume, A. G.; Hoffman, E. K.; Scott, R. W.; Wright, J. S.; Putcha, G. V.; Salvi, R. J. Targeted deletion of the cytosolic Cu/Zn-superoxide dismutase gene (Sod1) increases susceptibility to noise-induced hearing loss. *Audiol. Neurootol.* 1999; 4(5):237-46.

[29] Liu, Y. M.; Li, X. D.; Guo, X.; Liu, B.; Lin, A. H.; Rao, S. Q. Association between polymorphisms in SOD1 and noise-induced hearing loss in Chinese workers. *Acta Otolaryngol.* 2010;130(4):477-86.

[30] Wang, J.; Li, J.; Peng, K.; Fu, Z. Y.; Tang, J.; Yang, M. J.; Chen, Q.C.; Wang, J.; Li, J.; Peng, K.; Fu, Z. Y.; Tang, J.; Yang, M. J.; Chen, Q.C. Association of the C47T polymorphism in superoxide dismutase gene 2 with noise-induced hearing loss: a meta-analysis. *Braz. J. Otorhinolaryngol.*, 2017; 83(1):80-87.

[31] Dahn, H. C.; Benedetti, M. S.; Dostert, P. Differential changes in superoxide dismutase activity in brain and liver of old rats and mice. *J. Neurochem.*, 1983; 40:1003-1007.

[32] Dovrat, A.; Gershon, D. Rat lens superoxide dismutase and glucose-6-phosphate dehydrogenase: studies on the catalytic activity and the fate of enzyme antigen as a function of age. *Exp. Eye. Res.*, 1981; 33(6):651-61.

[33] Keithley, E. M.; Canto, C.; Zheng, Q. Y.; Wang, X.; Fischel-Ghodsian, N.; Johnson, K. R. Cu/Zn superoxide dismutase and age-related hearing loss. *Hear. Res.*, 2005; 209(1-2):76-85.

[34] Gagliardi, S.; Cova, E.; Davin, A.; Guareschi, S.; Abel, K.; Alvisi, E.; Laforenza, U.; Ghidoni, R.; Cashman, J. R.; Ceroni, M.; Cereda, C. SOD1 mRNA expression in sporadic amyotrophic lateral sclerosis. *Neurobiology of Disease*, 2010;39(2):198–203.

[35] Kitoh, R.; Nishio, S. Y.; Ogawa, K.; Okamoto, M.; Kitamura, K.; Gyo, K.; Sato, H.; Nakashima, T.; Fukuda, S.; Fukushima, K.; Hara, A.; Yamasoba, T.; Usami, S. SOD1 gene polymorphisms in sudden sensorineural hearing loss. *Acta Otolaryngol.*, 2016;136(5):465-9.

[36] Gyo, K. Experimental study of transient cochlear ischemia as a cause of sudden deafness. *World J. Otorhinolaryngol.*, 2013; 3(1):1-15.

[37] Teranishi, M.; Uchida, Y.; Nishio, N.; Kato, K.; Otake, H.; Yoshida, T.; Suzuki, H.; Sone, M.; Sugiura, S.; Ando, F.; Shimokata, H.; Nakashima, T. Polymorphisms in genes involved in oxidative stress response in patients with sudden sensorineural hearing loss and Ménière's disease in a Japanese population. *DNA Cell. Biol.*, 2012; 31(10):1555-62.

[38] Coling, D. E.; Yu, K. C.; Somand, D.; Satar, B.; Bai, U.; Huang, T. T.; Seidman, M. D.; Epstein, C. J.; Mhatre, A. N.; Lalwani, A. K. Effect of SOD1 overexpression on age- and noise-related hearing loss. *Free Radic. Biol. Med.*, 2003; 34(7):873-80.

[39] Kawamoto, K.; Sha, S. H.; Minoda, R.; Izumikawa, M.; Kuriyama, H.; Schacht, J.; Raphael, Y. Antioxidant gene therapy can protect hearing and hair cells from ototoxicity. *Mol. Ther.*, 2004; 9(2):173-81.

[40] Serra, A.; Maiolino, L.; Cocuzza, S.; Di Luca, M.; Campione, G.; Licciardello, L.; Di Mauro, P. Assessment of oxidative stress markers and hearing thresholds in patients with obstructive sleep apnea-hypopnoea treated with cysteine and superoxide dismutase therapy. *Acta Biomed.* 2017;87(3):253-258.

INDEX

A

B

C

D

E

I

K

L

M

T

U

V

W

X

Y

Z